Intelligent Systems Reference Library

Volume 195

Series Editors

Janusz Kacprzyk, Polish Academy of Sciences, Warsaw, Poland

Lakhmi C. Jain, Faculty of Engineering and Information Technology, Centre for
Artificial Intelligence, University of Technology, Sydney, NSW, Australia;
KES International, Shoreham-by-Sea, UK;
Liverpool Hope University, Liverpool, UK

The aim of this series is to publish a Reference Library, including novel advances and developments in all aspects of Intelligent Systems in an easily accessible and well structured form. The series includes reference works, handbooks, compendia, textbooks, well-structured monographs, dictionaries, and encyclopedias. It contains well integrated knowledge and current information in the field of Intelligent Systems. The series covers the theory, applications, and design methods of Intelligent Systems. Virtually all disciplines such as engineering, computer science, avionics, business, e-commerce, environment, healthcare, physics and life science are included. The list of topics spans all the areas of modern intelligent systems such as: Ambient intelligence, Computational intelligence, Social intelligence, Computational neuroscience, Artificial life, Virtual society, Cognitive systems, DNA and immunity-based systems, e-Learning and teaching, Human-centred computing and Machine ethics, Intelligent control, Intelligent data analysis, Knowledge-based paradigms, Knowledge management, Intelligent agents, Intelligent decision making, Intelligent network security, Interactive entertainment, Learning paradigms, Recommender systems, Robotics and Mechatronics including human-machine teaming, Self-organizing and adaptive systems, Soft computing including Neural systems, Fuzzy systems, Evolutionary computing and the Fusion of these paradigms, Perception and Vision, Web intelligence and Multimedia.

** Indexing: The books of this series are submitted to ISI Web of Science, SCOPUS, DBLP and Springerlink.

More information about this series at http://www.springer.com/series/8578

Erik Cuevas · Primitivo Diaz ·
Octavio Camarena

Metaheuristic Computation:
A Performance Perspective

 Springer

Erik Cuevas
CUCEI
Universidad de Guadalajara
Guadalajara, Mexico

Primitivo Diaz
CUCEI
Universidad de Guadalajara
Guadalajara, Mexico

Octavio Camarena
CUCEI
Universidad de Guadalajara
Guadalajara, Mexico

ISSN 1868-4394   ISSN 1868-4408   (electronic)
Intelligent Systems Reference Library
ISBN 978-3-030-58102-2   ISBN 978-3-030-58100-8   (eBook)
https://doi.org/10.1007/978-3-030-58100-8

This Springer imprint is published by the registered company Springer Nature Switzerland AG
The registered company address is: Gewerbestrasse 11, 6330 Cham, Switzerland

Preface

Many problems in engineering nowadays concern with the goal of an "optimal" solution. Several optimization methods have, therefore, emerged, being researched and applied extensively to different optimization problems.

Typically, optimization methods arising in engineering are computationally complex because they require evaluation of a quite complicated objective function, which is often multimodal, non-smooth or even discontinuous. The difficulties associated with using mathematical optimization on complex engineering problems have contributed to the development of alternative solutions. Metaheuristic computation techniques are stochastic optimization methods that have been developed to obtain near-optimum solutions in complex optimization problems, for which traditional mathematical techniques normally fail.

Metaheuristic methods use as inspiration our scientific understanding of biological, natural or social systems, which at some level of abstraction can be represented as optimization processes. In their operation, searcher agents emulate a group of biological or social entities that interact with each other based on specialized operators that model a determined biological or social behavior. These operators are applied to a population (or several sub-populations) of candidate solutions (individuals) that are evaluated with respect to their fitness. Thus, in the evolutionary process, individual positions are successively approximated to the optimal solution of the system to be solved.

Due to their robustness, metaheuristic techniques are well-suited options for industrial and real-world tasks. They do not need gradient information and they can operate on each kind of parameter space (continuous, discrete, combinatorial, or even mixed variants). Essentially, the credibility of evolutionary algorithms relies on their ability to solve difficult real-world problems with a minimal amount of human effort.

There exist some common features that clearly appear in most of the metaheuristic approaches, such as the use of diversification to force the exploration of regions of the search space, rarely visited until now, and the use of intensification or exploitation, to investigate some promising regions thoroughly. Another common feature is the use of memory to archive the best solutions encountered.

Metaheuristic schemes are used to estimate the solutions to complex optimization problems. They are often designed to meet the requirements of particular problems because no single optimization algorithm can solve all problems competitively. Therefore, in order to select an appropriate metaheuristic technique, its relative efficacy must be appropriately evaluated.

Metaheuristic search methods are so numerous and varied in terms of design and potential applications; however, for such an abundant family of optimization techniques, there seems to be a question that needs to be answered: Which part of the design in a metaheuristic algorithm contributes more to its better performance? One widely accepted principle among researchers considers that metaheuristic search methods can reach a better performance when an appropriate balance between exploration and exploitation of solutions is achieved. While there seems to exist a general agreement on this concept, in fact, there is barely a vague conception of what the balance of exploration and exploitation really represents. Indeed, the classification of search operators and strategies present in a metaheuristic method is often ambiguous, since they can contribute in some way to explore or exploit the search space.

Several works that compare the performance among metaheuristic approaches have been reported in the literature. Nevertheless, they suffer from one of the following limitations:

(A) Their conclusions are based on the performance of popular evolutionary approaches over a set of synthetic functions with exact solutions and well-known behaviors, without considering the application context or including recent developments. (B) Their conclusions consider only the comparison of their final results, which cannot evaluate the nature of a good or bad balance between exploration and exploitation.

Numerous books have been published taking into account many of the most widely known methods, namely simulated annealing, tabu search, evolutionary algorithms, ant colony algorithms, particle swarm optimization or differential evolution, but attempts to consider the discussion of alternative approaches is scarce. The excessive publication of developments based on the simple modification of popular metaheuristic methods presents an important disadvantage, in that it distracts attention away from other innovative ideas in the field of metaheuristics. There exist several alternative metaheuristic methods that consider very interesting concepts; however, they seem to have been completely overlooked in favor of the idea of modifying, hybridizing or restructuring traditional metaheuristic approaches.

The goal of this book is to present the performance comparison of various metaheuristic techniques when they face complex optimization problems. In the comparisons, the following criteria have been adopted:

(I) special attention is paid to recently developed metaheuristic algorithms,
(II) the balance between exploration and exploitation has been considered to evaluate the search performance and

(III) the use of demanding applications such as energy problems and leucocyte detection.

This book includes ten chapters. The book has been structured so that each chapter can be read independently from the others. Chapter 1 describes the main characteristics and properties of metaheuristic and swarm methods. This chapter analyses the most important concepts of metaheuristic and swarm schemes.

The first part of the book that involves Chaps. 2 and 3 presents recent metaheuristic algorithms, their operators and characteristics. In Chap. 2, a modification to the original LS algorithm referred to as LS-II, is presented. In LS-II, the locust motion model of the original algorithm is modified, incorporating the main characteristics of the new biological formulations. As a result, LS-II improves its original capacities of exploration and exploitation of the search space. In order to test its performance, the proposed LS-II method is compared against several state-of-the-art evolutionary methods considering a set of benchmark functions and engineering problems. Experimental results demonstrate the superior performance of the proposed approach in terms of solution quality and robustness.

Chapter 3 presents a methodology to implement human-knowledge-based optimization. In the scheme, a Takagi-Sugeno Fuzzy inference system is used to reproduce a specific search strategy generated by a human expert. Therefore, the number of rules and its configuration only depend on the expert experience without considering any learning rule process. Under these conditions, each fuzzy rule represents an expert observation that models the conditions under which candidate solutions are modified in order to reach the optimal location. To exhibit the performance and robustness of the proposed method, a comparison to other well-known optimization methods is conducted. The comparison considers several standard benchmark functions, which are typically found in the scientific literature. The results suggest a high performance of the proposed methodology.

The second part of the book which involves Chaps. 4–9 present the use of recent metaheuristic algorithms in different domains. The idea is to compare the potential of new metaheuristic alternatives algorithms from a practical perspective.

In Chap. 4, an improved version of the CSA method is presented to solve complex optimization problems of energy. In the new algorithm, two features of the original CSA are modified: (I) the awareness probability (AP) and (II) the random perturbation. With such adaptations, the new approach preserves solution diversity and improves the convergence to difficult high multimodal optima. In order to evaluate its performance, the proposed algorithm has been tested in a set of four optimization problems that involve induction motors and distribution networks. The results demonstrate the high performance of the proposed method when it is compared with other popular approaches.

Chapter 5 presents a nonlinear system identification method based on the Hammerstein model. In the proposed scheme, the system is modeled through the adaptation of an ANFIS scheme, taking advantage of the similarity between it and the Hammerstein model. To identify the parameters of the modeled system, the proposed approach uses a recent nature-inspired method called the Gravitational

Search Algorithm (GSA). Compared to most existing optimization algorithms, GSA delivers better performance in complex multimodal problems, avoiding critical flaws such as a premature convergence to sub-optimal solutions. To show the effectiveness of the proposed scheme, its modeling accuracy has been compared with other popular evolutionary computing algorithms through numerical simulations on different complex models.

In Chap. 6, the recently proposed States of Matter Search (SMS) metaheuristic optimization method is proposed for maximizing the average State of Charge of Plug-in Hybrid Electric Vehicles (PHEVs) within a charging station. In our experiments, several different scenarios consisting of different numbers of PHEVs were considered. To test the feasibility of the proposed approach, comparative experiments were performed against other popular PHEVs' State of Charge maximization approaches based on swarm optimization methods. The results obtained on our experimental setup show that the proposed SMS-based State of Charge (SoC) maximization approach has an outstanding performance in comparison to that of the other compared methods, and as such, proves to be superior for tackling the challenging problem of PHEVs' smart charging.

Chapter 7 uses the locust search (LS) method for solving the optimal capacitor placement (OCP) problem. The proposed approach has been tested by considering several IEEE's radial distribution test systems, and its performance has also been compared against that of other techniques currently reported on the literature to solve the OCP problem. The experimental results suggest that the proposed LS-based method is able to competitively solve the OCP problem in terms of accuracy and robustness.

In Chap. 8, an accurate methodology for retinal vessel and optic disc segmentation is presented. The proposed scheme combines two different techniques: The Lateral Inhibition (LI) and the Differential Evolution (DE). The LI scheme produces a new image with enhanced contrast between the background and retinal vessels. Then, the DE algorithm is used to obtain the appropriate threshold values through the minimization of the cross-entropy function from the enhanced image. To evaluate the performance of the proposed approach, several experiments over images extracted from STARE, DRIVE, and DRISHTI-GS databases have been conducted. Simulation results demonstrate high performance of the proposed scheme in comparison with similar methods reported in the literature.

Chapter 9 presents an algorithm for the automatic detection of white blood cells embedded into complicated and cluttered smear images that consider the complete process as a circle detection problem. This chapter illustrates the use of metaheuristic computation schemes for the automatic detection of white blood cells embedded into complicated and cluttered smear images. The approach considers the identification problem as the process of detection of multi-ellipse shapes. The scheme uses the Differential Evolution (DE) method, which is easy to use, maintains a quite simple computation scheme presenting acceptable convergence properties. The approach considers the use of five edge points as agents that represent the candidate ellipses in the edge image of the smear. A cost function assesses if such candidate ellipses are present in the actual edge image. With the values of the cost function, the set of agents is modified by using the DE algorithm so that they can approximate the white blood cells contained in the edge-only map of the image.

The third part, which includes Chap. 10, compares the metaheuristic methods considering the balance between the exploration and exploitation of their search mechanism.

Finally, Chap. 10 presents an experimental analysis that quantitatively evaluates the balance between exploration and exploitation on several of the most important and better-known metaheuristic algorithms. In the study, a dimension-wise diversity measurement is used to assess the balance of each scheme considering a representative set of 40 benchmark problems that involve multimodal, unimodal, composite and shifted functions. As a result, the analysis provides several observations that allow understanding of how this balance affects the results in each type of functions, and which balance is producing for better solutions.

As authors, we wish to thank many people who were somehow involved in the writing process of this book. We express our gratitude to Prof. Lakhmi C. Jain, who so warmly sustained this project. Acknowledgments also go to Dr. Thomas Ditzinger and Divya Meiyazhagan, who so kindly also support this book project.

Guadalajara, Mexico   Erik Cuevas
                                                                              Primitivo Diaz
                                                                           Octavio Camarena

Contents

Chapter 1
Introductory Concepts of Metaheuristic Computation

Abstract This chapter presents the main concepts of metaheuristic schemes. The objective of this chapter is to introduce the characteristics and properties of these approaches. An important propose of this chapter is also to recognize the importance of metaheuristic methods to solve optimization problems in the cases in which traditional techniques are not suitable.

1.1 Formulation of an Optimization Problem

Most of the industrial and engineering systems require the use of an optimization process for their operation. In such systems, it is necessary to find a specific solution that is considered the best in terms of a cost function. In general terms, an optimization scheme corresponds to a search strategy that has as an objective to obtain the best solution considering a set of potential alternatives. This bets solution represents the best possible solution that, according to the cost function, solves the optimization formulation appropriately [1].

Consider a public transportation system of a specific town, for illustration proposes. In this example, it is necessary to find the "best" path to a particular target destination. To assess each possible alternative and then get the best possible solution, an adequate criterion should be taken into account. A practical criterion could be the relative distances among all possible routes. Therefore, a hypothetical optimization scheme selects the option with the smallest distance as a final output. It is important to recognize that several evaluation elements are also possible, which could consider other important criteria such as the number of transfers, the time required to travel from a location to another or ticket price.

Optimization can be formulated as follows: Consider a function $f : S \rightarrow \Re$ which is called the cost function, find the argument that minimizes f:

$$x^* = \arg \min_{x \in S} f(\mathbf{x}) \tag{1.1}$$

© Springer Nature Switzerland AG 2021

E. Cuevas et al., *Metaheuristic Computation: A Performance Perspective*,
Intelligent Systems Reference Library 195,
https://doi.org/10.1007/978-3-030-58100-8_1

S corresponds to the search space that refers to all possible solutions. In general terms, each possible solution solves in a different quality the optimization problem. Commonly, the unknown elements of x represent the decision variables of the optimization formulation. The cost function f determines the quality of each candidate solution. It evaluates the way in which a candidate element x solves the optimization formulation

In the example of public transportation, S represents all subway stations, bus lines, etc., available in the database of the transportation system. $\mathbf{x}$ represents a possible path that links the start location with the final destination. $f(\mathbf{x})$ is the cost function that assesses the quality of each possible route. Some other constraints can be incorporated as a part of the problem definition, such as the ticket price or the distance to the destination (in different situations, it is taken into account the combination of both indexes, depending on our preferences).

When additional constraints exist, the optimization problem is called constrained optimization (different from unconstrained optimization where such restrictions are not considered). Under such conditions, an optimization formulation involves the next elements:

- One or several decision variables from $\mathbf{x}$, which integrate a candidate solution
- A cost function $f(\mathbf{x})$ that evaluates the quality of each solution $\mathbf{x}$
- A search space S that defines the set of all possible solutions to $f(\mathbf{x})$
- Constraints that represent several feasible regions of the search space S.

In practical terms, an optimization approach seeks within a search space S a solution for $f(\mathbf{x})$ in a reasonable period of time with enough accuracy. The performance of the optimization method also depends on the type of formulation. Therefore, an optimization problem is well-defined if the following conditions are established:

1. There is a solution or set of solutions that satisfy the optimal values.
2. There is a specific relationship between a solution and its position so that small displacements of the original values generate light deviations in the objective function $f(\mathbf{x})$.

1.2 Classical Optimization Methods

Once an engineering problem has been translated into a cost function, the next operation is to choose an adequate optimization method. Optimization schemes can be divided into two sets: classical approaches and metaheuristic methods [2].

Commonly, $f(\mathbf{x})$ presents a nonlinear association in terms of its modifiable decision variables $\mathbf{x}$. In classical optimization methods, an iterative algorithm is employed to analyze the search space efficiently. Among all approaches introduced in the literature, the methods that use derivative-descent principles are the most popular. Under such techniques, the new position x_{k+1} is determined from the current location x_k in a direction toward $\mathbf{d}$:

$$x_{k+1} = x_k + \alpha \mathbf{d}, \tag{1.2}$$

where α symbolizes the learning rate that determines the extent of the search step in the direction to $\mathbf{d}$. The direction $\mathbf{d}$ in Eq. (1.2) is computed, assuming the use of the gradient ($\mathbf{g}$) of the objective function $f(\cdot)$.

One of the most representative methods of the classical approaches is the steepest descent scheme. Due to simplicity, this technique allows us to solve efficiently objective functions. Several other derivative-based approaches consider this scheme as the basis for the construction of more sophisticated methods. The steepest descent scheme is defined under the following formulation:

$$x_{k+1} = x_k - \alpha \mathbf{g}(f(x)), \tag{1.3}$$

In spite of its simplicity, classical derivative-based optimization schemes can be employed as long as the cost function presents two important constraints:

- (I) The cost function can be two-timed derivable.
- (II) The cost function is unimodal; i.e., it presents only one optimal position.

The optimization problem defines a simple case of a derivable and unimodal objective function. This function fulfills the conditions (I) and (II):

$$f(x_1, x_2) = 10 - e^{-(x_1^2 + 3 \cdot x_2^2)} \tag{1.4}$$

Figure 1.1 shows the function defined by formulation (1.4).

Considering the current complexity in the design of systems, there are too few cases in which traditional methods can be applied. Most of the optimization problems imply situations that do not fulfill the constraints defined for the application of gradient-based methods. One example involves combinatorial problems where

Fig. 1.1 Cost function with unimodal characteristics

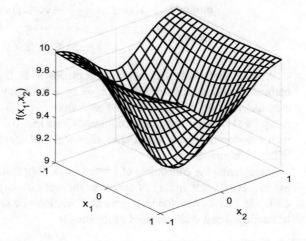

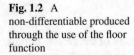

Fig. 1.2 A non-differentiable produced through the use of the floor function

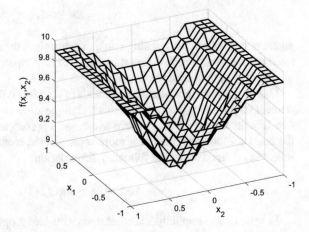

there is no definition of differentiation. There exist also many situations why an optimization problem could not be differentiable. One example is the "floor" function, which delivers the minimal integer number of its argument. This operation applied in Eq. (1.4) transforms the optimization problem from (Eq. 1.5) one that is differentiable to others not differentiable. This problem can be defined as follows (Fig. 1.2):

$$f(x_1, x_2) = \text{floor}\left(10 - e^{-(x_1^2 + 3 \cdot x_2^2)}\right) \tag{1.5}$$

Although an optimization problem can be differentiable, there exist other restrictions that can limit the use of classical optimization techniques. Such a restriction corresponds to the existence of only an optimal solution. This fact means that the cost function cannot present any other prominent local optima. Let us consider the minimization of the Griewank function as an example.

$$\text{minimize } f(x_1, x_2) = \frac{x_1^2 + x_2^2}{4000} - \cos(x_1)\cos\left(\frac{x_2}{\sqrt{2}}\right) + 1$$
$$\text{subject to} \qquad \begin{array}{c} -5 \leq x_1 \leq 5 \\ -5 \leq x_2 \leq 5 \end{array} \tag{1.6}$$

A close analysis of the formulation presented in Eq. (1.6); it is clear that the optimal global solution is located in $x_1 = x_2 = 0$. Figure 1.3 shows the cost function established in Eq. (1.6). As can be seen from Fig. 1.3, the cost function presents many local optimal solutions (multimodal) so that the gradient-based techniques with a randomly generated initial solution will prematurely converge to one of them with a high probability.

Considering the constraints of gradient-based approaches, it makes difficult their use to solve a great variety of optimization problems in engineering. Instead, some other alternatives which do not present restrictions are needed. Such techniques can be employed in a wide range of problems [3].

Fig. 1.3 The Griewank
function with multimodal
characteristics

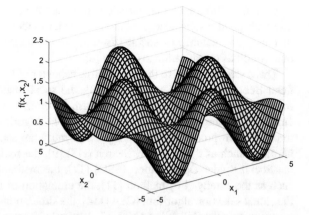

1.3   Metaheuristic Computation Schemes

Metaheuristic [4] schemes are derivative-free methods, which do not need that
the cost function maintains the restrictions of being two-timing differentiable or
unimodal. Under such conditions, metaheuristic methods represent global optimization methods which can deal with several types of optimization problems such as non-
convex, nonlinear, and multimodal problems subject to linear or nonlinear constraints
with continuous or discrete decision variables.

The area of metaheuristic computation maintains a rich history. With the demands
of more complex industrial processes, it is necessary for the development of new optimization techniques that do not require prior knowledge (hypotheses) on the optimization problem. This lack of assumptions is the main difference between classical
gradient-based methods. In fact, the majority of engineering system applications
are highly nonlinear or characterized by noisy objective functions. Furthermore,
in several cases, there is no explicit deterministic expression for the optimization
problem. Under such conditions, the evaluation of each candidate solution is carried
out through the result of an experimental or simulation process. In this context, the
metaheuristic methods have been proposed as optimization alternatives.

A metaheuristic approach is a generic search strategy used to solve optimization problems. It employs a cost function in an abstract way, considering only its
evaluations in particular positions without considering its mathematical properties.
Metaheuristic methods do not need any hypothesis on the optimization problem nor
any kind of prior knowledge on the objective function. They consider the optimization
formulation as "black boxes" [5]. This property is the most prominent and attractive
characteristic of metaheuristic computation.

Metaheuristic approaches collect the necessary knowledge about the structure of
an optimization problem by using the information provided by all solutions (i.e.,
candidate solutions) assessed during the optimization process. Then, this knowledge
is employed to build new candidate solutions. It is expected that these new solutions
present better quality than the previous ones.

Currently, different metaheuristic approaches have been introduced in the literature with good results. These methods consider modeling our scientific knowledge of biological, natural, or social systems, which, under some perspective, can be understood as optimization problems [6].

These schemes involve the cooperative behavior of bee colonies such as the Artificial Bee Colony (ABC) technique [8], the social behavior of bird flocking and fish schooling such as the Particle Swarm Optimization (PSO) algorithm [7], the emulation of the bat behavior such as the Bat Algorithm (BA) method [10], the improvisation process that occurs when a musician searches for a better state of harmony such as the Harmony Search (HS) [9], the social-spider behavior such as the Social Spider Optimization (SSO) [12], the mating behavior of firefly insects such as the Firefly (FF) method [11], the emulation of immunological systems as the clonal selection algorithm (CSA) [14], the simulation of the animal behavior in a group such as the Collective Animal Behavior [13], the emulation of the differential and conventional evolution in species such as the Differential Evolution (DE) [16], the simulation of the electromagnetism phenomenon as the electromagnetism-Like algorithm [15] and Genetic Algorithms (GA) [17], respectively.

1.3.1   Generic Structure of a Metaheuristic Method

In general terms, a metaheuristic scheme refers to a search strategy that emulates under a particular point of view a specific biological, natural or social system. A generic metaheuristic method involves the following characteristics:

1. Maintain a population of candidate solutions.
2. This population is dynamically modified through the production of new solutions.
3. A cost function associates the capacity of a solution to survive and reproduce similar elements.
4. Different operations are defined in order to explore an appropriately exploit the space of solutions through the production of new promising solutions.

Under the metaheuristic methodology, it is expected that, on average, candidate solutions enhance their quality during the evolution process (i.e., their ability to solve the optimization formulation). In the operation of the metaheuristic scheme, the operators defined in its structure will produce new solutions. The quality of such solutions will be improved as the number of iterations increases. Since the quality of each solution is associated with its capacity to solve the optimization problem, the metaheuristic method will guide the population towards the optimal global solution. This powerful mechanism has allowed the use of metaheuristic schemes to several complex engineering problems in different domains [18–20].

Most of the metaheuristic schemes have been devised to solve the problem of finding a global solution of a nonlinear optimization problem with box constraints in the following form:

Fig. 1.4 Generic procedure
of a metaheuristic scheme

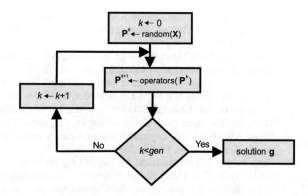

$$\text{Maximize/Minimize } f(\mathbf{x}), \quad \mathbf{x} = (x_1, \ldots, x_d) \in \Re^d$$
$$\text{subject to} \qquad \mathbf{x} \in \mathbf{X} \tag{1.7}$$

where $f : \Re^d \rightarrow \Re$ represents a nonlinear function whereas $\mathbf{X} = \{\mathbf{x} \in \Re^d | l_i \le x_i \le u_i, i = 1, \ldots, d\}$ corresponds to the feasible search space, restriced by the lower (l_i) and upper (u_i) bounds.

With the objective of solving the problem of Eq. (1.6), under the metaheuristic computation methodology, a group (population) $\mathbf{P}^k (\{\mathbf{p}_1^k, \mathbf{p}_2^k, \ldots, \mathbf{p}_N^k\})$ of N possible solutions (individuals) is modified from a start point ($k = 0$) to a total *gen* number iterations ($k = gen$). In the beginning, the scheme starts initializing the set of N candidate solutions with random values uniformly distributed between the pre-specified lower (l_i) and upper (u_i) limits. At each generation, a group operations are used over the current population $\mathbf{P}^k$ to generate a new set of individuals $\mathbf{P}^{k+1}$. Each possible solution $\mathbf{p}_i^k$ ($i \in [1, \ldots, N]$) corresponds to a d-dimensional vector $\{p_{i,1}^k, p_{i,2}^k, \ldots, p_{i,d}^k\}$ where each element represents a decision variable of the optimization problem to be solved. The capacity of each possible solution $\mathbf{p}_i^k$ to solve the optimization problem is assessed by considering a cost function $f(\mathbf{p}_i^k)$ whose delivered value symbolizes the fitness value of $\mathbf{p}_i^k$. As the evolution process progresses, the best solution $\mathbf{g}$ ($g_1, g_2, \ldots g_d$) seen so-far is maintained since it is the best available solution. Figure 1.4 shows an illustration of the generic procedure of a metaheuristic method.

References

1. Akay, B., Karaboga, D.: A survey on the applications of artificial bee colony in signal, image, and video processing. SIViP **9**(4), 967–990 (2015)
2. Yang, X.-S.: Engineering Optimization. Wiley (2010)
3. Treiber, M.A.: Optimization for Computer Vision An Introduction to Core Concepts and Methods. Springer (2013)
4. Dan S.: Evolutionary Optimization Algorithms. Wiley (2013)

5. Blum, C., Roli, A.: Metaheuristics in combinatorial optimization: overview and conceptual comparison. ACM Comput. Surveys (CSUR) **35**(3), 268–308 (2003). https://doi.org/10.1145/937503.937505

6. Nanda, S.J., Panda, G.: A survey on nature inspired metaheuristic algorithms for partitional clustering. Swarm Evolution. Computat. **16**, 1–18 (2014)

7. Kennedy, J., Eberhart, R.: Particle swarm optimization. In: Proceedings of the 1995 IEEE International Conference on Neural Networks, vol. 4, pp. 1942–1948 (1995)

8. Karaboga, D.: An idea based on honey bee swarm for numerical optimization. Technical Report-TR06. Engineering Faculty, Computer Engineering Department, Erciyes University (2005)

9. Geem, Z.W., Kim, J.H., Loganathan, G.V.: A new heuristic optimization algorithm: harmony search. Simulations **76**, 60–68 (2001)

10. Yang, X.S.: A new metaheuristic bat-inspired algorithm. In: Cruz, C., González, J., Krasnogor, G.T.N., Pelta, D.A. (eds.) Nature Inspired Cooperative Strategies for Optimization (NISCO 2010), Studies in Computational Intelligence, vol. 284, pp. 65–74. Springer Verlag, Berlin (2010)

11. Yang, X.S.: Firefly algorithms for multimodal optimization, in: stochastic algorithms: foundations and applications. SAGA 2009, Lecture Notes in Computer Sciences, vol. 5792, pp. 169–178 (2009)

12. Cuevas, E., Cienfuegos, M., Zaldívar, D., Pérez-Cisneros, M.: A swarm optimization algorithm inspired in the behavior of the social-spider. Expert Syst. Appl. **40**(16), 6374–6384 (2013)

13. Cuevas, E., González, M., Zaldivar, D., Pérez-Cisneros, M., García, G.: An algorithm for global optimization inspired by collective animal behaviour. Discre. Dyn. Nat. Soc. 2012, art. no. 638275

14. de Castro, L.N., von Zuben, F.J.: Learning and optimization using the clonal selection principle. IEEE Trans. Evol. Comput. **6**(3), 239–251 (2002)

15. Birbil, Ş.I., Fang, S.C.: An electromagnetism-like mechanism for global optimization. J. Glob. Optim. **25**(1), 263–282 (2003)

16. Storn, R., Price, K.: Differential evolution-a simple and efficient adaptive scheme for global optimisation over continuous spaces. Technical ReportTR-95–012, ICSI, Berkeley, CA (1995)

17. Goldberg, D.E.: Genetic algorithm in search optimization and machine learning. Addison-Wesley (1989)

18. Cuevas, E.: Block-matching algorithm based on harmony search optimization for motion estimation. Appl. Intell. **39**(1), 165–183 (2013)

19. Díaz-Cortés, M.-A., Ortega-Sánchez, N., Hinojosa, S., Cuevas, E., Rojas, R., Demin, A.: A multi-level thresholding method for breast thermograms analysis using Dragonfly algorithm. Infrared Phys. Technol. **93**, 346–361 (2018)

20. Díaz, P., Pérez-Cisneros, M., Cuevas, E., Hinojosa, S., Zaldivar, D.: An improved crow search algorithm applied to energy problems. Energies **11**(3), 571 (2018)

Chapter 2
An Enhanced Swarm Method Based on the Locust Search Algorithm

Abstract In the evolutionary methods, the optimal balance of exploration and exploitation performance in search strategies improve efficiency to found the best solution. In this chapter, an improved swarm optimization technique called Locust search II (LS-II) based on the desert locust swarm behavior, adapted to an emulation of a group of locusts which interacts to each other based on the biological laws of the cooperative swarm is proposed for solving global optimization problems. Such methodology combines a technique of exploration which avoids premature convergence effectively and a technique of exploitation able to intensify the global solutions. The proposed LS-II method was tested over several well-known benchmark test functions and engineering optimization problems and its performance was further compared against those of other state-of the-art methods such as Particle Swarm Optimization (PSO), Artificial Bee Colony (ABC), Bat Algorithm (BA), Differential Evolution (DE), Harmony Search (HS) and the original Locust Search (LS). Our experimental results show LS-II to be superior to all other compared methods in terms of solution quality and as such proves to be an excellent alternative to handle complex optimization problems.

2.1 Introduction

For the last few decades, optimization approaches inspired by the natural collective behavior of insects and animals have captivated the attention of many researchers. These techniques, commonly referred to as swarm optimization methods, combine deterministic rules and randomness with the purpose of mimicking some kind of natural phenomena, typically manifested in the form of a swarm behavior. Search strategies based in swarm behaviors have demonstrated to be adequate to solve complex optimization problems, often delivering significantly better solutions than those produced by traditional methods. Currently, an extensive variety of swarm-based optimization techniques can be found on the literature. Some examples include Particle Swarm Optimization (PSO), which emulates the social behavior of flocking birds or fishes [1], the Artificial Bee Colony (ABC) approach, which considers the

© Springer Nature Switzerland AG 2021
E. Cuevas et al., *Metaheuristic Computation: A Performance Perspective*,
Intelligent Systems Reference Library 195,
https://doi.org/10.1007/978-3-030-58100-8_2

cooperative behavior manifested in bee colonies [2], the Cuckoo Search (CS) algorithm, which simulates the brood parasitism behavior manifested by cuckoo birds
[3], the Firefly Algorithm (FA), which mimics he distinctive bioluminescence-based
behavior observed in fireflies [4], among others.

The Locust Search (LS) [5] algorithm is a swarm optimization approach inspired
in the biological behavior of the desert locusts (*Schistocerca gregaria*). Biologically,
locusts experiment two opposite phases: solitary and social. In the solitary phase,
locusts avoid contact with others conspecifics in order to explore promising food
sources. In opposition, in the social phase, locusts frantically aggregate around abundant foods sources (such as plantations) devastating them. This aggregation is carried
on through the attraction to those elements that are found the best food sources. By
integrating these two distinctive behaviors, LS maintains powerful global and local
search capacities which enable it to solve effectively a wide range of complex optimization problems such as image processing [6], parameter estimation of chaotic
systems [7], pattern recognition [8], among others.

In spite of its interesting characteristics, LS has some shortcomings from the
evolutionary computing point of view. In its social phase operator, LS performs a
series of random walks around some of the best individuals with the objective of
refining its original solution. This process seems to be appropriate for the purpose of
local exploitation; however, the excessive production of candidate solutions adversely
increases the algorithm computational overload. Other important flaws of LS are the
switch between the solitary and social phase. In LS, both behaviors are performed
in the same cycle; this means that each individual search agent in LS is subject
to performing both global and local movements at each iteration of the algorithms
search process. Although this mechanism could be useful in some contexts, it is
known that a misbalance between exploration and exploitation of solutions is able
to significantly degrade any algorithm's performance, thus making this approach
somewhat unreliable.

Base on this premise, in this chapter, a modification to the original LS is proposed
in order to better and more efficiently handle global optimization problems. In our
modified approach, coined Locust Search II (LS-II), a probabilistic criterion, is introduced in order to control how the solitary and social phase operators are addressed by
the algorithm, essentially allowing the swarm of locusts to "decide" when to apply
each behavioral phase. Also, individual decision making based on probabilities is
introduced to the algorithm's social phase operator as a mean to allow search agents
to manifest an attraction toward prominent individuals within the population. Said
modifications were devised with the purpose of providing a better balance between
the exploration and exploitation of solutions, while also allowing an important reduction on the algorithms overall computational cost. To demonstrate the proficiency of
our proposed approach, we performed a series of experiments over a set of benchmark
test functions. The results of our approach are compared to those produced by the
original LS, as well as some other state-of-the-art optimization techniques, including
Particle Swarm Optimization (PSO) [1], Artificial Bee Colony (ABC) [2], Bat Algorithm (BA) [9], Differential Evolution (DE) [10], and Harmony Search (HS) [11].
Also, in order to enhance the analysis of our proposed method, we also performed

comparative experiments over several popular engineering optimization problems, including the design of pressure vessels, gear trains [12], tension/compression springs [13], welded beams [14], three-bar truss [15], parameter estimation for FM synthesizers [16], and Optimal Capacitor Placement for Radial Distribution Networks [17, 18]. The experimental results for both sets shows that LS-II are superior not only over the original LS, but also over all other compared methods in terms of solution quality and robustness.

2.2 The Locust Search Algorithm

Locust Search (LS) [5] is a global optimization algorithm based on the gregarious behavior observed in swarms of desert locusts [19–24]. In LS, search agents are represented by a set of N individual locusts $\mathbf{L} = \{\mathbf{l}_1, \mathbf{l}_2, \ldots, \mathbf{l}_N\}$ (with N representing the total population size), interacting with each other while moving through a n-dimensional feasible solution space. Each individual position $\mathbf{l}_i = [l_{i,1}, l_{i,2}, \ldots, l_{i,n}]$ is defined within a bounded space $\mathbf{S} = \{\mathbf{x} \in \mathbb{R}^n | lb_d \leq x_d \leq ub_d\}$ ($with \mathbf{x} = [x_1, x_2, \ldots, x_d]$ and where lb_d and ub_d represent the lower and upper bounds at the d-th dimension, respectively) and represents a candidate solution for a specified optimization problem.

Similar to other swarm-based optimization techniques, LS comprises an iterative scheme in which search agents change their positions at each generation of the algorithm during its evolution. The change of position applied to each individual is conducted by a set of operators inspired in the two behavioral phases observed in desert locusts: solitary phase and social phase.

2.2.1 LS Solitary Phase

In the solitary phase, individuals move in different locations searching for promising food sources (solutions) while they avoid to aggregate with other conspecifics. This scheme is modeled by considering attraction and repulsion forces manifested among individuals within the population. Therefore, for any iteration "k", the total attraction and repulsions forces (collectively referred to as social force) experienced by a specific individual "i" are given by the following expression:

$$\mathbf{S}_i^k = \sum_{\substack{j=1 \\ i \neq 1}}^{N} \mathbf{s}_{ij}^k \tag{2.1}$$

where $\mathbf{s}_{ij}^k$ denotes the pairwise attraction (or repulsion) between locust 'i' and some other individual 'j', and is given by:

$$\mathbf{s}_{ij}^k = \rho\left(\mathbf{l}_i^k, \mathbf{l}_j^k\right) \cdot s\left(r_{ij}^k\right) \cdot \mathbf{d}_{ij} + \mathrm{rand}(1, -1) \tag{2.2}$$

where $r_{ij}^k = \mathbf{l}_i^k - \mathbf{l}_j^k$ denotes the Euclidian distance between the locusts 'i' and 'j'. Therefore, $\mathbf{d}_{ij} = \left(\mathbf{l}_j^k - \mathbf{l}_i^k\right)/r_{ij}^k$ stands for the unit vector pointing from $\mathbf{l}_i^k$ to $\mathbf{l}_j^k$, while $\mathrm{rand}(1, -1)$ is a random number drawn from the uniform distribution of $[-1, 1]$. Furthermore, the value $s\left(r_{ij}^k\right)$ represents the so-called social factor and is given by:

$$s\left(r_{ij}^k\right) = F \cdot e^{-\frac{r_{ij}^k}{L}} - e^{-r_{ij}^k} \tag{2.3}$$

where the user-defined parameters F and L denotes the attraction/repulsion magnitude and attractive length scale, respectively. Finally, the operator $\rho\left(\mathbf{l}_i^k, \mathbf{l}_j^k\right)$ is known as dominance value. To apply this operator, it is first assumed that each locust $\mathbf{l}_i^k \in \mathbf{L}^k$ $(\mathbf{l}_1^k, \mathbf{l}_2^k, \ldots, \mathbf{l}_N^k)$ is ranked with a number from 0 (best individual) to $N - 1$ (worst individual) depending on their respective fitness value. With that being said, the dominance value may be given as follows:

$$\rho\left(\mathbf{l}_i^k, \mathbf{l}_j^k\right) = \begin{cases} e^{-\left(\mathrm{rank}\left(l_i^k\right)/N\right)} & \text{if rank}\left(\mathbf{l}_i^k\right) \leq \text{ rank}\left(\mathbf{l}_j^k\right) \\ e^{-\left(\mathrm{rank}\left(l_j^k\right)/N\right)} & \text{if rank}\left(\mathbf{l}_i^k\right) > \text{ rank}\left(\mathbf{l}_j^k\right) \end{cases} \tag{2.4}$$

As a result of the influence of total social force $\mathbf{S}_i^k$, each individual locust 'i' manifest a certain tendency to move toward (or away from) other members within the locust swarm. As such, the new position adopted by locust 'i' as a result of $\mathbf{S}_i^k$ may be expressed as:

$$\mathbf{l}_i^* = \mathbf{l}_i^k + \mathbf{S}_i^k \tag{2.5}$$

The result of applying the solitary movement operators to each individual $\mathbf{l}_k^* \in \mathbf{L}^k$ is a new set of candidate solutions $\mathbf{L}^* = \left\{\mathbf{l}_1^*, \mathbf{l}_2^*, \ldots, \mathbf{l}_N^*\right\}$, which represent the positions taken by each individual as a result of the influence exerted by all other members in the swarm.

2.2.2 LS Social Phase

The social phase operation is applied to refine some of the best candidate solutions $\mathbf{L}^* = \left\{\mathbf{l}_1^*, \mathbf{l}_2^*, \ldots, \mathbf{l}_N^*\right\}$ generated by applying the solitary phase movement operator described in Sect. 2.2.1. For this purpose, a subset of solutions $\mathbf{B} = \left\{\mathbf{b}_1, \ldots, \mathbf{b}_q\right\}$, comprised by the q best solutions within the set $\mathbf{L}^*$, is first defined. Then, for each solution $\mathbf{l}_1^* \in \mathbf{B}$, a set of h random solutions $Mi = \{\mathbf{m}_1^i, ..., \mathbf{m}_h^i\}$ is generated within a corresponding subspace $Ci \in \mathbf{S}$, whose limits are given by

$$C_{i,n}^{lower} = b_{i,n} - r$$
$$C_{i,n}^{upper} = b_{i,n} + r \qquad (2.6)$$

where $C_{i,n}^{lower}$ and $C_{i,n}^{upper}$ represent the upper and lower bounds of each subspace C_i at the n-th dimension, respectively, while $b_{i,n}$ stands for the n-th element (decision variable) from solution $\mathbf{b}_i$, and where

$$r = \frac{\sum_{n=1}^{d}\left(b_n^{upper} - b_n^{lower}\right)}{d} \cdot \beta \qquad (2.7)$$

with b lower d and b upper d denoting the lower and upper bounds at the n-th dimension, respectively, whereas d stands for total number of decision variables. Furthermore, $\beta \in [0, 1]$ represents a scalar factor which modulates the size of Ci. Finally, the best solution from $\mathbf{l}_i^* \in \mathbf{B}$ and its respective h random solutions $(\mathbf{m}_1^i, \mathbf{m}_2^i, ..., \mathbf{m}_h^i)$ is assigned as the position for individual "i" at the following iteration "$k + 1$". That is,

$$\mathbf{l}_i^{k+1} = \text{best}\left(\mathbf{l}_i^*, \mathbf{m}_1^i, \mathbf{m}_2^i, \ldots, \mathbf{m}_h^i\right) \qquad (2.8)$$

It is worth nothing that any solution $\mathbf{l}_i^*$ that is not grouped within the set of best solutions $\mathbf{B}$ is excluded by the social phase operator. As such, the final position update applied to each individual 'i' within the whole swarm may be summarized as follows:

$$\mathbf{l}_i^{k+1} = \begin{cases} best\left(\mathbf{l}_i^*, \mathbf{m}_1^i, \mathbf{m}_2^i, \ldots, \mathbf{m}_h^i\right) if \, \mathbf{l}_i^* \in B \\ \mathbf{l}_i^* \, if \, \mathbf{l}_i^* \notin B \end{cases} \qquad (2.9)$$

2.3 The LS-II Algorithm

The LS algorithm proposes an interesting global and local search capacities in the form of the so called solitary and social phase operators. In spite of its interesting characteristics, LS has some shortcomings from an evolutionary computing point of view. Specifically, in its social phase operator, LS performs a series of random walks around some of the best individuals with the objective of refining their current solutions; while this process seems to be appropriate for the purpose of local exploitation, excessive evaluations of candidate solutions are known to adversely increase the algorithm computational load, a phenomenon that is undesirable to virtually any optimization approach. Another important flaw of the original LS approach is given by the way in which the solitary and social phase behavioral operators are addressed. In LS, both the solitary and social phases operators are coded to be performed in sequence at each iteration of the algorithms search process; this means that each individual is subject to undergone to both global and local operations without considering its individual decision. Although this mechanism could be useful in some contexts, it produces a misbalance between exploration and exploitation which could significantly degrade its performance.

In this chapter a modification to the original LS algorithm, referred to as LS-II, is proposed. In our approach, instead of applying both the solitary and social phase operators in the same iteration, we propose a simple probabilistic scheme to selectively apply either one of this two behaviors at each iteration of the algorithms search process. Furthermore, we have also proposed some changes the original LS's social operators: essentially, instead of performing a series of local evaluations around promising solutions, LS-II's social operator incorporates a probabilistic selection mechanism that allows locusts to be attracted toward some other prominent individuals within the population. This modification not only allows to reduce the algorithms overall computational cost, but also enhances LS-II's convergence speed.

2.3.1 Selecting Between Solitary and Social Phases

The first modification, considered in the original LS, involves the selection for any given iteration between the solitary phase and the social phase. In the proposed LS-II approach, we assume that, at the beginning of the search process, individuals have a tendency to perform intensive exploration over the feasible search space by means of their solitary phase behavior, whereas in later stages of the evolution process exploitation is favored by the social phase behavior. Also, it is considered that the individuals maintain a certain probability to behave as either solitary or social depending on the current stage (iteration) of the algorithm's search process. Therefore, as the evolutionary process progresses, the probability that an element behaves in a social manner increases, while the probability that it acts in solitary behavior, is reduced. Under such circumstances, at each iteration "k", the behavioral phase $P(\mathbf{L}^k)$ applied to the population $\mathbf{L}^k$ is chosen as follows:

$$P(\mathbf{L}^k) = \begin{cases} \textbf{Solitary} \text{ if } rand \leq p^k \\ \textbf{Social} \text{ if } rand > p^k \end{cases} \qquad (2.10)$$

where $rand$ denotes a random number sampled within the uniformly distributed interval $[0, 1]$, while value p^k, called behavior probability, is given by

$$p^k = 1 - \frac{k}{itern} \tag{2.11}$$

where *itern* stands for the maximum number of iterations considered in the search process. From the expression, it is clear that, as the iterative process increases, the value of the behavior probability *pk* is linearly decreased. Intuitively, other monotonically decreasing models (such as logarithmic or exponential functions) could also be considered to handle this value.

It is important to remark that from both operators (solitary and social) only the social phase has been modified, and the solitary phase behavior proposed by the original LS remains unchanged. In Fig. 2.1, we show a comparison between the original LS algorithm and the proposed LS-II approach.

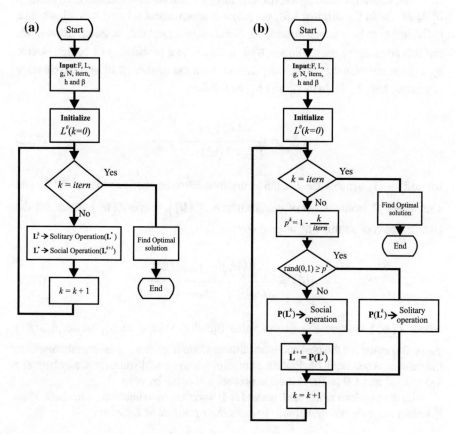

Fig. 2.1 Flowchart of **a** LS and **b** LS-II

2.3.2 Modified Social Phase Operator

The second modification considers several changes to the original LS algorithm's social phase, each aimed at giving every individual $\mathbf{l}_i^k$ the ability to move toward potentially good solution within the feasible solution space instead of performing a series of local function evaluations. This mechanism does not only allow taking advantage of the information provided by all individuals within the population, but also allows a significant reduction in computational load by taking away excessive and unnecessary function evaluations. Similarly to the original LS approach, the LS-II's social phase starts by selecting a subset of solutions $\mathbf{B}^k = \{\mathbf{b}_1^k, ..., \mathbf{b}_q^k\}$ that include the q best solutions from the total set $\mathbf{L}^k = \{\mathbf{l}_1^k, ..., \mathbf{l}_q^k\}$ (therefore, $\mathbf{B}^k \subseteq \mathbf{L}^k$). However, instead of defining a set of random candidate solutions around each position $\mathbf{b}_j^k \subseteq \mathbf{B}^k$ (as in the original LS), we propose a movement scheme that allows each individual $\mathbf{l}_i^k$ to be attracted toward the direction of a randomly selected solution $\mathbf{b}_j^k$. For this purpose, to each solution $\mathbf{b}_j^k$ is assigned to a probability of being selected by a given individual $\mathbf{l}_i^k$, which depends on both the quality of $\mathbf{b}_j^k$ and the distance separating both individuals ($\mathbf{l}_j^k$ and $\mathbf{b}_j^k$) as follows:

$$P_{\mathbf{l}_i\mathbf{b}_j}^k = \frac{A\left(\mathbf{b}_j^k\right) * e^{-\mathbf{l}_i^k - \mathbf{b}_j^k}}{\sum_{n=1}^{q} A\left(\mathbf{b}_n^k\right) * e^{-\mathbf{l}_i^k - \mathbf{b}_n^k}} \tag{2.12}$$

where $\mathbf{l}_i^k - \mathbf{b}_j^k$ denotes the Euclidian distance between the individual "i" ($\mathbf{l}_i^k$) and a member "j" from the set of best solutions $\mathbf{B}^k\left(\mathbf{b}_j^k\right)$, while $A\left(\mathbf{b}_j^k\right)$ stands for the attractiveness of solution $\mathbf{b}_j^k$ as given by

$$A\left(\mathbf{b}_j^k\right) = \frac{f\left(\mathbf{b}_j^k\right) - f_{worst}\left(\mathbf{B}^k\right)}{f_{best}\left(\mathbf{b}^k\right) - f_{worst}\left(\mathbf{B}^k\right) + \in} \tag{2.13}$$

where $f\left(\mathbf{b}_j^k\right)$ denotes the fitness value (quality) related to $\mathbf{b}_j^k$, while $f_{best}\left(\mathbf{B}^k\right)$ $f_{worst}\left(\mathbf{B}^k\right)$ stand for the best and worst fitness value from among the members within the set of best solutions $\mathbf{B}^k$. Finally, ε corresponds to a small value (typically between 1.0×10^{-4} and 1.0×10^{-5}) used to prevent a division by cero.

With the previous being said, under LS-II's social phase behavior, each individual $\mathbf{l}_1^k$ within the swarm population L updates their position as follows:

$$\mathbf{l}_i^{k+1} = \mathbf{l}_i^k + 2 \cdot \left(\mathbf{b}_r^k - \mathbf{l}_i^k\right) \cdot rand \tag{2.14}$$

Fig. 2.2 LS-II's modified social phase operator. During the social phase, each individual locust $\mathbf{l}_i$ moves toward the direction of a randomly chosen member $\mathbf{b}_r$ from among the q best individuals $(\mathbf{b}_1, \ldots, \mathbf{b}_q)$ within the whole locust population. The figure illustrates the case for $q = 3$

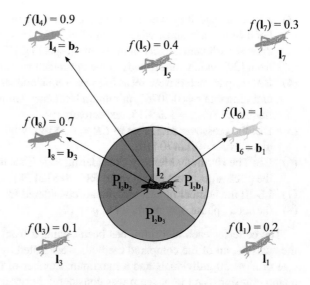

where $\mathbf{b}_r^k$ (with $r \in [1, \ldots, q]$) is a randomly chosen solution $\mathbf{b}_j^k \in \mathbf{B}^k$, selected by applying the roulette selection method [25] with regard to their respective probabilities P_{ij}^k (relative to $\mathbf{l}_i^k$ and $\mathbf{b}_j^k$), while $rand$ stands for a random number drawn from within the uniformly distributed interval $[0, 1]$ (see Fig. 2.2).

2.4 Experiments and Results

2.4.1 Benchmark Test Functions

In our experiments, we studied LS-II's performance with regard to the optimization of 13 well-known benchmark tests functions (see Appendix A) operated in 30 dimensions. Furthermore, we have also compared out proposed method's result against those produced by other popular optimization techniques, including Particle Swarm Optimization (PSO) [1], Artificial Bee Colony (ABC) [2], Bat Algorithm (BA) [9], Differential Evolution (DE) [10], Harmony Search (HS) [11], and the original Locust Search (LS) [5].

The parameters setting for each of the methods has been performed by using the automatic algorithm configuration software known as *iRace* [26]. This method allows obtaining the best possible parameter configuration under a specific set of benchmark problems. After the *iRace* analysis, the following settings have been found:

(1) LS: the algorithm's parameters were set to $q = 10$, $h = 3$ [5].
(2) PSO: the cognitive and social coefficients are set to $c1 = 1.9460$ and $c2 = 1.8180$, respectively. Also, the inertia weight factor ω is set to decrease linearly from 0.543254 to 0.33362 as the search process evolves [1].

(3) ABC: the algorithm was implemented by setting the parameter *limit* = num Of Food Sources * dims, where num of *FoodSources* = N (population size) and dims = n (dimensionality of the solution space) [2]. The code has been obtained from [27] which corresponds to the optimized version (v2).

(4) BA: the parameters were set as follows: initial loudness rate $A = 0.311895$, pulse emission rate $r = 0.70767$, minimum frequency $f$min = 0.3272, and maximum frequency $f$max = 1.65815, respectively [9].

(5) DE: the crossover rate is set to $CR = 0.58143$, while the differential weight is given as $F = 0.12470$ [10].

(6) HS: The Harmony Memory Considering Rate is set to $HCMR = 0.98950$ while the pitch adjustment rate is set to $PA = 0.812174$ [11].

(7) LS-II: the number of best individuals considered for

(8) the social phase operator is set to $q = 10$.

The code for each algorithm has been obtained from their original sources. In the analysis, all of the compared methods were tested by considering a population size of $N = 50$ individuals and a maximum number of iterations $itern = 1000$ as a stop criterion. Said tests setup was considered in order to keep consistency with other similar works currently found in the literature. All experiments were performed on MatLAB® R2016a, running on a computer with an Intel® Core™ i7-3.40 GHz processor, and Windows 8 (64-bit, 8 GB of memory) as operating system.

Our experimental setup aims to compare the performance of LS-II against those of PSO, ABC, BA, DE, HS, and LS. The averaged experimental results, corresponding to 30 individual runs, are reported in Table 2.1 where the best outcomes for each test function are boldfaced. The results reported in this chapter consider the following performance indexes: Average Best fitness (AB), Median Best fitness (MB), and the standard deviation of the best finesses (SD). As shown by our experiments, LS-II demonstrates to be superior not only to the original LS approach, but also to all other of the compared methods. Such a notorious performance is related to both the excellent global search capabilities provided LSII's solitary phase operators, the highly effective exploitation mechanism of our modified social phase operators, and naturally the better trade-off between exploration and exploitation that is achieved through the selective application of said behavioral phases. Furthermore, Wilcoxon's rank sum test for independent samples [28] was conducted over the best fitness values found by each of the compared methods on 30 independent test runs (30 executions by each set). Table 2.2 reports the p-values produced by said test for the pairwise comparison over two independent sets of samples (LS-II versus LS, ABC, BA, DE, and HS), by considering a 5% significance level. For the Wilcoxon test, a null hypothesis assumes that there is a significant difference between the mean values of two algorithms. On the other hand, an alternative hypothesis (rejection of the null hypothesis) suggests that the difference between the mean values of both compared methods is insignificant. As shown by all of the p-values reported in Table 2.3, enough evidence is provided to reject the null hypothesis (all values are less than 0.05, and as such satisfy the 5% significance level criteria). This proves the statistical significant

Table 2.1 Minimization results for the benchmark test functions illustrated in Appendix A2

		PSO	DE	ABC	HS	BA	LS	LS-II
f_1	f_{mean}	1.50×10^{02}	4.01×10^{-13}	2.41×10^{-01}	5.25×10^{02}	7.58×10^{01}	4.32×10^{-04}	$\mathbf{4.33 \times 10^{-32}}$
	f_{median}	9.35×10^{-08}	3.25×10^{-13}	2.17×10^{-01}	5.38×10^{02}	5.72×10^{01}	2.17×10^{-04}	$\mathbf{1.02 \times 10^{-33}}$
	f_{std}	3.41×10^{02}	2.25×10^{-13}	1.40×10^{-01}	7.60×10^{01}	6.77×10^{01}	6.52×10^{-04}	$\mathbf{9.30 \times 10^{-32}}$
f_2	f_{mean}	1.06×10^{01}	2.48×10^{-08}	7.93×10^{-03}	9.96×10^{00}	2.92×10^{-01}	5.12×10^{-02}	$\mathbf{2.06 \times 10^{-16}}$
	f_{median}	7.68×10^{00}	2.50×10^{-08}	7.52×10^{-03}	1.03×10^{01}	1.59×10^{-02}	5.50×10^{-02}	$\mathbf{1.69 \times 10^{-16}}$
	f_{std}	7.95×10^{00}	7.24×10^{-09}	1.72×10^{-03}	1.49×10^{00}	4.65×10^{-01}	1.71×10^{-02}	$\mathbf{2.13 \times 10^{-16}}$
f_3	f_{mean}	1.65×10^{04}	8.60×10^{-12}	3.11×10^{00}	1.22×10^{04}	1.02×10^{01}	1.82×10^{-03}	$\mathbf{3.86 \times 10^{-30}}$
	f_{median}	1.49×10^{04}	7.83×10^{-12}	2.93×10^{00}	1.21×10^{04}	7.13×10^{-03}	1.55×10^{-03}	$\mathbf{1.28 \times 10^{-31}}$
	f_{std}	1.07×10^{04}	3.92×10^{-12}	1.14×10^{00}	2.41×10^{03}	3.35×10^{01}	1.38×10^{-03}	$\mathbf{8.39 \times 10^{-30}}$
f_4	f_{mean}	1.59×10^{-01}	1.95×10^{-02}	6.36×10^{-01}	5.78×10^{-04}	5.10×10^{-04}	1.05×10^{-02}	$\mathbf{7.43 \times 10^{-17}}$
	f_{median}	1.60×10^{-01}	1.93×10^{-02}	6.32×10^{-01}	5.85×10^{-04}	5.31×10^{-04}	1.15×10^{-02}	$\mathbf{5.12 \times 10^{-17}}$
	f_{std}	4.15×10^{-02}	2.44×10^{-03}	3.67×10^{-02}	3.46×10^{-02}	6.78×10^{-05}	6.63×10^{-03}	$\mathbf{7.49 \times 10^{-17}}$
f_5	f_{mean}	2.49×10^{02}	2.60×10^{01}	1.55×10^{03}	4.00×10^{03}	1.35×10^{00}	4.27×10^{02}	$\mathbf{2.89 \times 10^{01}}$
	f_{median}	8.03×10^{01}	2.60×10^{01}	1.51×10^{03}	3.84×10^{03}	4.62×10^{-01}	3.74×10^{02}	$\mathbf{2.90 \times 10^{01}}$
	f_{std}	4.22×10^{02}	8.64×10^{-01}	5.30×10^{02}	1.11×10^{03}	3.16×10^{00}	2.81×10^{02}	$\mathbf{4.15 \times 10^{-02}}$
f_6	f_{mean}	6.67×10^{00}	$\mathbf{0.00 \times 10^{00}}$	3.33×10^{-02}	5.80×10^{01}	3.00×10^{01}	5.58×10^{-02}	6.06×10^{00}
	f_{median}	0.00×10^{00}	$\mathbf{0.00 \times 10^{00}}$	0.00×10^{00}	5.70×10^{01}	2.90×10^{01}	5.67×10^{-02}	5.89×10^{00}
	f_{std}	2.54×10^{01}	$\mathbf{0.00 \times 10^{00}}$	1.83×10^{-01}	1.26×10^{01}	1.53×10^{01}	1.07×10^{-02}	5.16×10^{-01}
f_7	f_{mean}	3.68×10^{04}	1.04×10^{01}	1.95×10^{02}	3.80×10^{03}	4.86×10^{02}	2.21×10^{-02}	$\mathbf{8.77 \times 10^{-03}}$
	f_{median}	3.00×10^{04}	1.06×10^{01}	1.96×10^{02}	3.64×10^{03}	4.83×10^{02}	1.90×10^{-02}	$\mathbf{2.18 \times 10^{-03}}$
	f_{std}	1.89×10^{04}	5.00×10^{-01}	8.38×10^{01}	1.17×10^{03}	1.80×10^{02}	1.12×10^{-02}	$\mathbf{9.65 \times 10^{-03}}$

(continued)

Table 2.1 (continued)

		PSO	DE	ABC	HS	BA	LS	LS-II
f_8	f_{mean}	1.10×10^{04}	1.07×10^{04}	-5.03×10^{143}	1.09×10^{04}	2.94×10^{03}	-8.61×10^{05}	$\mathbf{-2.59 \times 10^{03}}$
	f_{median}	1.10×10^{04}	1.07×10^{04}	-1.24×10^{139}	1.09×10^{04}	2.66×10^{03}	-4.43×10^{05}	$\mathbf{-2.62 \times 10^{03}}$
	f_{std}	9.75×10^{01}	1.31×10^{01}	2.67×10^{144}	4.37×10^{01}	1.45×10^{03}	1.06×10^{06}	$\mathbf{2.80 \times 10^{02}}$
f_9	f_{mean}	1.40×10^{03}	6.55×10^{01}	2.77×10^{02}	5.86×10^{03}	-2.51×10^{03}	2.69×10^{-02}	$\mathbf{0.00 \times 10^{00}}$
	f_{median}	6.97×10^{01}	6.47×10^{01}	2.70×10^{02}	5.59×10^{03}	-2.11×10^{03}	3.44×10^{-03}	$\mathbf{0.00 \times 10^{00}}$
	f_{std}	3.46×10^{03}	5.58×10^{00}	1.65×10^{01}	1.08×10^{03}	4.98×10^{02}	4.00×10^{-02}	$\mathbf{0.00 \times 10^{00}}$
f_{10}	f_{mean}	2.00×10^{01}	2.00×10^{01}	2.10×10^{01}	2.03×10^{01}	2.00×10^{01}	3.00×10^{-02}	$\mathbf{4.44 \times 10^{-15}}$
	f_{median}	2.00×10^{01}	2.00×10^{01}	2.10×10^{01}	2.03×10^{01}	2.00×10^{01}	1.73×10^{-02}	$\mathbf{4.44 \times 10^{-15}}$
	f_{std}	0.00×10^{00}	0.00×10^{00}	4.94×10^{-02}	4.74×10^{-02}	8.28×10^{-07}	3.56×10^{-02}	$\mathbf{0.00 \times 10^{00}}$
f_{11}	f_{mean}	1.28×10^{-01}	2.87×10^{-13}	4.55×10^{-01}	1.13×10^{00}	1.39×10^{-01}	1.42×10^{-04}	$\mathbf{0.00 \times 10^{00}}$
	f_{median}	1.23×10^{-02}	1.15×10^{-13}	4.38×10^{-01}	1.13×10^{00}	9.85×10^{-02}	1.49×10^{-05}	$\mathbf{0.00 \times 10^{00}}$
	f_{std}	3.18×10^{-01}	6.02×10^{-13}	1.59×10^{-01}	2.61×10^{-02}	1.81×10^{-01}	3.69×10^{-04}	$\mathbf{0.00 \times 10^{00}}$
f_{12}	f_{mean}	8.43×10^{01}	7.13×10^{01}	1.33×10^{07}	1.17×10^{06}	1.37×10^{05}	$\mathbf{5.08 \times 10^{-03}}$	1.34×10^{00}
	f_{median}	8.31×10^{01}	7.10×10^{01}	1.26×10^{07}	1.16×10^{06}	5.98×10^{04}	$\mathbf{5.12 \times 10^{-03}}$	1.32×10^{00}
	f_{std}	5.14×10^{00}	8.21×10^{-01}	2.47×10^{06}	4.01×10^{05}	1.89×10^{05}	$\mathbf{4.17 \times 10^{-04}}$	1.34×10^{-01}
f_{13}	f_{mean}	1.69×10^{05}	1.06×10^{02}	4.01×10^{05}	1.93×10^{04}	3.81×10^{01}	$\mathbf{1.08 \times 10^{-01}}$	3.15×10^{00}
	f_{median}	1.88×10^{05}	1.06×10^{02}	4.06×10^{05}	1.86×10^{04}	3.86×10^{01}	$\mathbf{1.05 \times 10^{-01}}$	3.00×10^{00}
	f_{std}	1.20×10^{05}	9.41×10^{00}	1.04×10^{05}	6.97×10^{03}	3.54×10^{00}	$\mathbf{1.37 \times 10^{-02}}$	2.69×10^{-01}

Results were averaged from 30 individual runs, each by considering population size of $N = 50$ and maximum number of iterations $itern = 1000$

Table 2.2 Wilcoxon's test comparison for LS-II versus PSO, DE, ABC, HS, BA and LS

Function	LS-II versus PSO	LS-II versus DE	LS-II versus ABC	LS-II versus HS	LS-II versus BA	LS-II versus LS
f_1	1.41×10^{-12}	2.02×10^{-11}	8.26×10^{-09}	7.48×10^{-12}	1.01×10^{-11}	4.86×10^{-12}
f_2	2.07×10^{-13}	1.58×10^{-13}	$2.26 \times 10+$	1.13×10^{-13}	7.25×10^{-13}	2.02×10^{-13}
f_3	1.47×10^{-11}	1.33×10^{-11}	3.97×10^{-11}	1.46×10^{-11}	1.42×10^{-11}	2.89×10^{-11}
f_4	2.21×10^{-12}	3.99×10^{-11}	1.65×10^{-08}	1.46×10^{-11}	1.80×10^{-11}	9.11×10^{-12}
f_5	3.41×10^{-12}	5.99×10^{-11}	2.48×10^{-08}	2.20×10^{-11}	2.74×10^{-11}	1.38×10^{-11}
f_6	4.62×10^{-12}	7.99×10^{-11}	3.31×10^{-08}	2.94×10^{-11}	3.68×10^{-11}	1.84×10^{-11}
f_7	5.83×10^{-12}	9.99×10^{-11}	4.13×10^{-08}	3.67×10^{-11}	4.61×10^{-11}	2.31×10^{-11}
f_8	7.03×10^{-12}	1.20×10^{-10}	4.96×10^{-08}	4.41×10^{-11}	5.55×10^{-11}	2.77×10^{-11}
f_9	8.24×10^{-12}	1.40×10^{-10}	5.79×10^{-08}	5.15×10^{-11}	6.49×10^{-11}	3.24×10^{-11}
f_{10}	9.45×10^{-12}	1.60×10^{-10}	6.61×10^{-08}	5.88×10^{-11}	7.42×10^{-11}	3.71×10^{-11}
f_{11}	1.07×10^{-11}	1.80×10^{-10}	7.44×10^{-08}	6.62×10^{-11}	8.36×10^{-11}	4.17×10^{-11}
f_{12}	1.19×10^{-11}	2.00×10^{-10}	8.26×10^{-08}	7.36×10^{-11}	9.30×10^{-11}	4.64×10^{-11}
f_{13}	1.31×10^{-11}	2.20×10^{-10}	9.09×10^{-08}	8.09×10^{-11}	1.02×10^{-10}	5.10×10^{-11}

The table shows the resulting p-values for each pair-wise comparison

Table 2.3 Minimization results for the engineering optimization problems described in Appendix B

		PSO	DE	ABC	HS	BA	LS	LS-II
J_1	J_{worst}	7302.944	6590.18	6497.539	9702.135	400595	28910.08	**6059.73**
	J_{best}	6876.679	6877.065	6138.001	6523.004	6683.271	6882.137	**6059.72**
	J_{avg}	6876.679	6897.291	6311.334	7210.724	103291.8	7986.061	**6059.72**
	J_{std}	5.61×10^{02}	1.33×10^{02}	7.84×10^{01}	6.82×10^{02}	1.20×10^{05}	5.87×10^{03}	**$4.30 \times 10\text{-}03$**
J_2	J_{worst}	1.17×10^{-01}	1.18×10^{-02}	3.85×10^{-02}	3.82×10^{-02}	5.14×10^{-02}	6.47×10^{-02}	**$9.61 \times 10\text{-}03$**
	J_{best}	1.61×10^{-02}	8.13×10^{-03}	2.50×10^{-02}	2.70×10^{-02}	7.99×10^{-02}	7.27×10^{-03}	**$5.39 \times 10\text{-}03$**
	J_{avg}	1.17×10^{-02}	7.96×10^{-03}	2.21×10^{-02}	1.09×10^{-02}	5.52×10^{-02}	7.03×10^{-02}	**$6.74 \times 10\text{-}03$**
	J_{std}	4.43×10^{05}	7.94×10^{-04}	1.31×10^{-18}	4.79×10^{05}	2.40×10^{05}	1.16×10^{-03}	**$1.03 \times 10\text{-}03$**
J_3	J_{worst}	1.9851	1.7511	1.7498	1.8012	1.7955	1.7449	**1.72485**
	J_{best}	1.7421	1.725	1.7249	1.7248	1.7249	1.7277	**1.72485**
	J_{avg}	1.7578	1.7298	1.7251	1.7562	1.7252	1.7311	**1.72485**
	J_{std}	7.41×10^{-03}	9.87×10^{-02}	3.50×10^{-04}	4.40×10^{-04}	2.50×10^{-03}	8.50×10^{-01}	**$7.16 \times 10\text{-}07$**
J_4	J_{worst}	347.763	347.2367	346.4102	347.5836	300.9182	320.2528	**279.936**
	J_{best}	346.4523	346.4538	346.4102	345.3665	284.377	301.3875	**279.725**
	J_{avg}	346.675	346.7444	346.4102	346.345	286.777	308.2089	**279.743**
	J_{std}	2.90×10^{-01}	1.98×10^{-01}	**5.74×10^{-14}**	1.98×10^{-01}	3.44×10^{00}	6.68×10^{-01}	4.12×10^{-02}
J_5	J_{worst}	28.51646	26.27407	27.60109	28.81973	30.08068	28.34164	**25.7868**
	J_{best}	25.5127	24.56246	22.85514	23.59037	22.32354	21.05672	**20.1103**
	J_{avg}	26.15824	25.82232	24.45296	25.59217	27.00966	25.42715	**23.898**
	J_{std}	2.19	6.46	3.50	1.72	2.14	2.55	**1.56**
J_6	J_{worst}	1.35×10^{-06}	9.09×10^{-07}	4.70×10^{-07}	1.79×10^{-06}	3.13×10^{-08}	1.57×10^{-02}	**$3.96 \times 10\text{-}08$**
	J_{best}	2.22×10^{-06}	1.78×10^{-09}	1.33×10^{-10}	2.67×10^{-09}	8.90×10^{-10}	4.45×10^{-07}	**$5.10 \times 10\text{-}13$**
	J_{avg}	7.17×10^{-03}	5.73×10^{-03}	4.30×10^{-03}	8.60×10^{-03}	2.87×10^{-03}	1.43×10^{-03}	**$1.27 \times 10\text{-}10$**
	J_{std}	2.06×10^{-03}	1.65×10^{-03}	1.23×10^{-03}	2.47×10^{-03}	8.23×10^{-05}	4.12×10^{-03}	**$7.77 \times 10\text{-}09$**

Results were obtained from 30 individual runs, each by considering population size of $N = 50$ and maximum number of iterations $itern = 1000$

of our proposed method's results and thus discard the possibility of them being a product of coincidence.

The analysis of the final fitness values cannot absolutely characterize the capacities of an optimization algorithm. Consequently, a convergence experiment on the seven compared algorithms has been conducted. The objective of this test is to evaluate the velocity with which an optimization scheme reaches the optimum. In the experiment, the performance of each algorithm is analyzed over a representative set of functions, including f_1, f_3, f_9, and f_{11}. In the test, all functions are operated considering 30 dimensions. In order to build the convergence graphs, we employ the raw data generated in the simulations. As each function is executed 30 times for each algorithm, we select the convergence data of the run which represents the median final result. Figure 2.3 presents the fitness evolution of each compared method for functions f_1, f_3, f_9, and f_{11}. According to Fig. 2.3, the LS-II method presents the overall best convergence when compared to the other compared methods; once again, this performance is testimony of LS-II's better trade-off between diversification and intensification, as well as the decision-making scheme incorporated to the algorithms social phase operators.

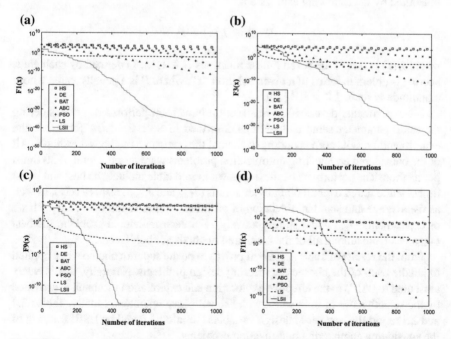

Fig. 2.3 Evolution curves of the run which represents the median final result. **a** $f1$, **b** $f3$, **c** $f9$, and **d** $f11$ represent the convergence graphs obtained by applying HS, DE, BA, ABC, PSO, LS, and LS-II. For each experiment, a population size of $N = 50$ and maximum number of iterations $itern = 1000$ have been considered

2.4.2 Engineering Optimization Problems

Metaheuristic methods have demonstrated their effectivity when they face complex engendering problems [30–32]. In order to further evaluate the performance of our proposed approach, we tested LS-II over several well-studied engineering design optimization problems, namely, the design of pressure vessels, gear trains [12], tension/compression springs [13] welded beams [14], three-bar trusses [15], parameter estimation for FM synthesizers [16], and Optimal Capacitor Placement for Radial Distribution Networks [17, 18]. Appendix B provides a detailed description of each engineering problem used in this study.

Similarly to the experiments reported in Sect. 4.1, we compared LS-II's performance over the previously mentioned real problems with those of Particle Swarm Optimization (PSO) [1], Artificial Bee Colony (ABC) [2], Bat Algorithm (BA) [9], Differential evolution (DE) [10], Harmony Search (HS) [11], and the original Locust Search (LS) [5]. For each design problem a penalty function is implemented to penalize the fitness value of solutions that infringe any of the restrictions modeled on each optimization task. The penalty implemented of the fitness function may be illustrated by the following expression:

$$J_i^*(\mathbf{x}) = J_i(\mathbf{x}) + P$$

where $Ji(\mathbf{x})$ is one of the seven fitness functions ($i \in 1, ..., 7$) defined by each engineering problem in terms of a given solution "x", where P is a penalty value whose magnitude is set to 1.0×10^5 [29].

All experiments described in this section have been performed by considering the same parameter setup and settings described in Sect. 2.4.1. In Table 2.3, the experimental results corresponding to each of the compared methods are shown. All of the considered engineering optimization problems were tested a total of 30 times per method. The performance indexes shown in said table include the best and worst fitness value found over all 30 individual runs (J_{best} and J_{worst}, respectively), as well as the average and standard deviation of all fitness values found over said each set of experiments (J_{avg} and J_{std}, respectively). For each individual problem, the best outcomes from among all of the compared methods are boldfaced.

From the results of Table 2.3, LS-II proves to be the superior choice when applied to handle each of the chosen engineering design problems. Naturally, this is easily attributed to LSII's to the effective exploration and exploitation mechanism provided by the selective use of both solitary and social phase operators. Finally, Tables 2.4 and 2.5 report the best set of design variables (solutions) found by LS-II for each of the considered engineering optimization problems.

Table 2.4 Best parameter configurations obtained by LS-II for each of the engineering optimization problems illustrated in Appendices B.2.1–B.2.6

Problem	x_1	x_2	x_3	x_4	x_5	x_6	$J_i(\mathbf{x})$
J_1	0.812500	0.4375001	42.09844	176.6365	–	–	6059.716
J_2	40	25	13	56	–	–	5.10463×10^{-13}
J_3	0.050664	0.329604	4.372104	–	–	–	0.0053911
J_4	0.2057303	3.4704743	9.0366238	0.205729	–	–	1.7248519
J_5	0.8696133	0.2169257	-7.2121×10^{-06}	-1.700119	-0.29988	–	279.72502
J_6	0.419345	2.422595	4.930669	2.403819	4.247489	4.90245	20.11029

Table 2.5 Best parameter configurations obtained by LS-II for the Optimal Capacitor Placement for the IEEE's 69-Bus Radial Distribution Networks illustrated in Appendix B.2.7

Problem	$x_1 - x_{15}$	x_{16}	$x_{17} - x_{59}$	x_{60}	$x_{61} - x_{69}$	$J_i(\mathbf{x})$
J_1	0	350	0	1200	0	83708.21

2.5 Conclusions

Locust Search (LS) is a swarm-based optimization method inspired in the natural behavior of the desert locust [5]. When it was first proposed, the LS method considered as basis the biological models extracted from observations of locust behaviors. In these models, the procedure to change from the solitary phase to the gregarious phase, together with the mechanism to select the element to which a locust will be attracted, is fixed and does not consider the decision of each locust in the behavioral process.

Currently, new computer vision experiments in insect tracking methods have conducted to the development of more accurate locust motion models than those produced by simple behavior observations. The most distinctive characteristic of such new models is the use of probabilities to emulate the locust decision process.

In this chapter, a modification to the original LS algorithm, referred to as LS-II, has been proposed to better handle global optimization problems. In LS-II, the locust motion model of the original algorithm is modified incorporating the main characteristics of the new biological formulations. Under the new model, probability decisions are considered to emulate the behavior of the locust colony. Therefore, the decisions, to change of behavioral phase and to select the attraction elements, are probabilistically taken.

The proposed LS-II method has been compared against several state-of-the-art evolutionary methods considering a set of benchmark functions and engineering problems. Experimental results demonstrate the superior performance of the proposed approach in terms of solution quality and robustness.

Appendix A

In Table 2.1, all of the 13 benchmark test functions that were implemented in our experiments are shown.

See Table 2.6.

Appendix B

In this section, we offer a detailed description for all of the engineering optimization problems considered in our experiments.

B2.1 Pressure Vessel Design Problem

As described in [12], the pressure vessel design problem consist on a constrained optimization problem in which the main objective is to minimize the cost of said artifact. This is usually achieved by defining an appropriate set of four design variables $\mathbf{x} = \left[x_{1}, x_{2}, x_{3}, x_{4} \right]$, namely: the shell's thickness T_{s} (x_{1}), head's thickness T_{h} (x_{2}), vessel's internal radius R (x_{3}) and vessel's length L (x_{4}) (see Fig. 2.4). With that being said, such cost minimization problem may be more formally expressed as follows.

Problem B2.1 (Pressure Vessel Design) Minimize:

$$J_1(\mathbf{x}) = 0.6224x_1x_2x_4 + 1.7781x_2x_3^2 + 3.1661x_1^2x_4 + 19.84x_1^2x_3$$

Subject to:

$$g_1(\mathbf{x}) = -x_1 + 0.0193x_3 \leq 0$$

$$g_2(\mathbf{x}) = -x_2 + 0.00954x_3 \leq 0$$

$$g_3(\mathbf{x}) = -\pi x_3^2x_4 - (4/3)\pi x_3^2 + 1{,}296{,}000 < 0$$

$$g_4(\mathbf{x}) = x_4 - 240 \leq 0$$

$$0 \leq x_i \leq 100, i = 1, 2$$

$$0 \leq x_i \leq 200, i = 3, 4$$

Table 2.6 Test functions used for our experiments

f_i	Name	Function	S	n
f_1	Sphere	$f(x) = \sum\limits_{i=1}^{n} x_i^2$ $x^* = (0, \ldots, 0); f(x^*) = 0$	$[-100, 100]^n$	30
f_2	Schewefel 2.22	$f(x) = \sum\limits_{i=1}^{n} |x_i| + \prod\limits_{i=1}^{n} |x_i|$ $x^* = (0, \ldots, 0); f(x^*) = 0$	$[-10, 10]^n$	30
f_3	Schewefel 1.2	$f(x) = \sum\limits_{i=1}^{n} \left(\sum\limits_{j=1}^{n} x_j \right)^2$ $x^* = (0, \ldots, 0); f(x^*) = 0$	$[-100, 100]^n$	30
f_4	Schewefel 2.22	$f(x) = \max\{|x|, 1 \leq i \leq n\}$ $x^* = (0, \ldots, 0); f(x^*) = 0$	$[-100, 100]^n$	30
f_5	Rosenbrock	$f(x) = \sum\limits_{i=1}^{n-1} \left[100(x_{i+1} - x_i^2)^2 + (x_i - 1)^2 \right]$ $x^* = (0, \ldots, 0); f(x^*) = 0$	$[-30, 30]^n$	30
f_6	Step	$f(x) = \sum\limits_{i=1}^{n} (|x_i| + 0.5|)^2$ $x^* = (0, \ldots, 0); f(x^*) = 0$	$[-100, 100]^n$	30
f_7	Quartic	$f(x) = \sum\limits_{i=1}^{n} i x_i^4 + rand(0, 1)$ $x^* = (0, \ldots, 0); f(x^*) = 0$	$[-1.28, 128]^n$	30

(continued)

Table 2.6 (continued)

f_i	Name	Function	S	n
f_8	Schewefel	$f(x) = \sum_{i=1}^{n} -x_i \sin(\sqrt{\|x_i\|})$ $x^* = (420,\ldots,420);\ f(x^*) = -418.9829 \times n$	$[-500, 500]^n$	30
f_9	Rastring	$f(x) = \sum_{i=1}^{n} [x_i^2 - 10\cos(2\pi x_i) + 10]$ $x^* = (0,\ldots,0);\ f(x^*) = 0$	$[-5.12, 5.12]^n$	30
f_{10}	Ackley's	$f(x) = -20\exp\left(-0.2\sqrt{\frac{1}{n}\sum_{i=1}^{n} x_i^2}\right)$ $-\exp\left(\frac{1}{n}\sum_{i=1}^{n}\cos(2\pi x_i)\right) + 20 + e$ $x^* = (0,\ldots,0);\ f(x^*) = 0$	$[-32, 32]^n$	30
f_{11}	Grienwank	$f(x) = \frac{1}{4000}\sum_{i=1}^{n} x_i^2 - \prod_{i=1}^{n}\cos\left(\frac{x_i}{\sqrt{i}}\right) + 1$ $x^* = (0,\ldots,0);\ f(x^*) = 0$	$[-600, 600]^n$	30
f_{12}	Penalized	$f(x) = \frac{\pi}{n}\left\{10\sin(\pi y_1) + \sum_{i=1}^{n}(y_i - 1)^2[1 + 10sin^2(\pi y_{i+1})] + (y_n - 1)^2\right\} + \sum_{i=1}^{n} u(x_i, 10, 100, 4)$ $y_i = 1 + \frac{x_i+1}{4}$ $u(x_i; a, k, m) = \begin{cases} k(x_i - a)^m & x_i > a \\ 0 & -a < x_i < a \\ k(-x_i - a)^m & x_i < -a \end{cases}$ $x^* = (0,\ldots,0);\ f(x^*) = 0$	$[-50, 50]^n$	30

(continued)

Table 2.6 (continued)

f_i	Name	Function	S	n
f_{13}	Penalized 2	$f(x) = 0.1\left\{ sin^2(3\pi x_1) + \sum_{i=1}^{n}(x_i-1)^2\big[1+sin^2(3\pi x_i)\big] + (x_n-1)^2\big[1+sin^2(2\pi x_n)\big] \right\} + \sum_{i=1}^{n} u(x_i, 5, 100, 4)$ $$u(x_i, a, k, m) = \begin{cases} k(x_i-a)^m & x_i > a \\ 0 & -a < x_i < a \\ k(-x_i-a)^m & x_i < -a \end{cases}$$ $x^* = (0, \ldots, 0); f(x^*) = 0$	$[-50, 50]^n$	30

In the table, **S** indicates the subset of Rn which comprises the function's search space and n indicates the function's dimension. Also, the value $f(\mathbf{x}^*)$ indicates the optimum value of each function, while $\mathbf{x}^*$ indicates the optimal solution

Fig. 2.4 Pressure vessel's
design variables

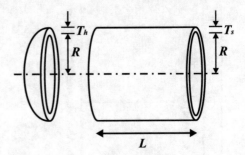

B2.2 Gear Train Design Problem

The optimization problem related to the design of gear trains (as the one shown
in Fig. 2.5) is to minimize the squared difference between the gear's teeth ratio and
some scalar value (1/6.931 in this case) [12]. As such, the decision variables comprise
the number of teeth on each of the gears A, B, D, and F. By considering $x_1 = A$,
$x_2 = B$, $x_3 = D$ and $x_4 = F$, the gear train design problem may then be formulated
as follows.

Problem B2.2 (Gear Train Design)
Minimize:

$$J_2(\mathbf{x}) = \left(\left(\frac{1}{6.931} \right) - \left(\frac{x_3 x_2}{x_1 x_4} \right) \right)^2$$

Subject to:

$$0 \leq x_i \leq 600, \, i = 1, 2, 3, 4$$

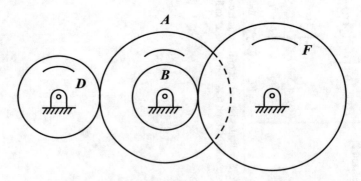

Fig. 2.5 Gear train design variables

B2.3 Tension/Compression Spring Design Problem

For this design problem the objective is to minimize the tension (or compression) experiment by a spring when subjected to some load P. This is usually achieved by optimizing with regard to three decision variables: the wire diameter d (x_1), the coil diameter D (x_2), and the number of active coils n (x_3). As such, the problem may be formulated as.

Problem B2.3 (Tension/Compression Spring Design)
Minimize:

$$J_3(\mathbf{x}) = (x_3 + 2)x_2 x_1^2$$

Subject to:

$$g_1(\mathbf{x}) = 1 - \frac{x_2^3 x_3}{71,785 x_1^4} \le 0$$

$$g_2(\mathbf{x}) = \frac{4x_2^2 - x_1 x_2}{12,566(x_2 x_1^3 - x_1^4)} + \frac{1}{5,108 x_1^2} - 1 \le 0$$

$$g_3(\mathbf{x}) = 1 - \frac{140.45 x_1}{x_2^2 + x_3} \le 0$$

$$g_4(\mathbf{x}) = \frac{x_1 + x_2}{1.5} \le 0$$

$$0.5 \le x_1 \le 2$$

$$0.25 \le x_2 \le 1.3$$

$$2 \le x_3 \le 15$$

See Fig. 2.6.

B2.4 Three-Bar Truss Design Problem

The objective behind the design of a symmetric three-bar truss (as shown in Fig. 2.7) is to minimize its fabrication cost while also allowing it to sustain a certain static load P and some stress σ. In this case, the design variables are comprised by the cross-sectional areas x_1, x_2 and x_3 corresponding to each of the truss's three bars. By

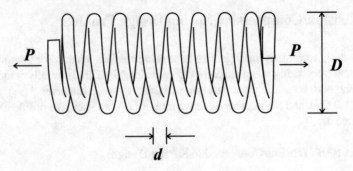

Fig. 2.6 The tension/compression spring design variables

Fig. 2.7 Three-bar truss
design variables

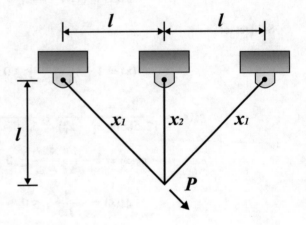

considering $x_3 = x_1$ (due to symmetry on the truss design), the optimization problem
may be then defined as follows

Problem B2.4 (Three-Bar Truss Design)
Minimize:

$$J_4(\mathbf{x}) = l\left(2\sqrt{2x_1} + x_2\right)$$

Subject to:

$$g_1(\mathbf{x}) = \frac{\sqrt{2x_1} + x_2}{\sqrt{2x_1^2 + 2x_1x_2}}P - \sigma \leq 0$$

$$g_2(\mathbf{x}) = \frac{x_2}{\sqrt{2x_1^2 + 2x_1x_2}}P - \sigma \leq 0$$

$$g_3(\mathbf{x}) = \frac{1}{\sqrt{2x_2^2 + x_1}} P - \sigma \le 0$$

$$0 \le x_i \le 1, i = 1, 2$$

$$l = 100 \text{ cm}, P = 2 \text{ kN/cm}^2, \sigma = 2 \text{ kN/cm}^2$$

B2.5 Welded Beam Design Problem

For this design problem, the aim is to minimize the fabrication cost of a beam meant to be welded to another surface. Said welded beam must be designed so that it is able to sustain certain amounts of shear stress (τ), bending stress (σ), bucking load (P_c) and deflection (δ). The design variables implied on the design of such a mechanism include the weld's thickness and length h and l, respectively, as well as the beam's width and thickness t and b, respectively (see Fig. 2.8). By considering $x_1 = h$, $x_2 = l$, $x_3 = t$, and $x_4 = b$, this design problem may then be formulated as.
Minimize:

$$J_5(\mathbf{x}) = 1.10471x_1^2 x_2 + 0.04811x_3 x_4(14 + x_2)$$

Subject to:

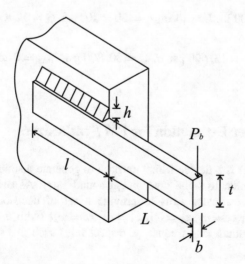

Fig. 2.8 Welded beam design variables

$$g_1(\mathbf{x}) = \tau(\mathbf{x}) - \tau_{\max} \geq 0$$

$$g_2(\mathbf{x}) = \sigma(\mathbf{x}) - \sigma_{\max} \leq 0$$

$$g_3(\mathbf{x}) = x_1 - x_4 \leq 0$$

$$g_4(\mathbf{x}) = 0.1047x_1^2 + 0.04811x_3x_4(14 + x_2) - 5 \leq 0$$

$$g_5(\mathbf{x}) = 0.125 - x_1 \leq 0$$

$$g_6(\mathbf{x}) = \delta(\mathbf{x}) - \delta_{\max} \leq 0$$

$$g_7(\mathbf{x}) = P - P_c(\mathbf{x})(?)$$

$$0.1 \leq x_i \leq 2, i = 1, 4$$

$$0.1 \leq x_i \leq 10, i = 2, 3$$

$$\sigma(\mathbf{x}) = \frac{6PL}{x_4x_3}, P_c(\mathbf{x}) = \frac{4.013E\sqrt{\frac{x_3^2x_4^6}{36}}}{L^2}\left(1 - \frac{x_3}{2L}\sqrt{\frac{E}{4G}}\right)$$

$$P = 6000 \text{ lb}, L = 14 \text{ in}, E = 30 \times 10^6 \text{ psi}, G = 12 \times 10^6 \text{ psi}$$

$$\tau_{\max} = 13,600 \text{ psi}, \sigma_{\max} = 30,000 \text{ psi}, \delta_{\max} = 0.25 \text{ in}$$

B2.6. Parameter Estimation for FM Synthesizers

An FM synthesizer is a device which purpose to generate a sound (signal) $y(\mathbf{x}, t)$ such that it is similar to some other target sound $y_0(t)$. As such, a FM synthesizer's parameter estimator aims to provide a set of decision variables $\mathbf{x} = [x_1 = a_1, x_2 = \omega_1, x_3 = a_2, x_4 = \omega_2, x_5 = a_3, x_6 = \omega_3]$ (with a_i and ω_i denoting a finite wave amplitude and frequency, respectively) such that the error between signals $y(\mathbf{x}, t)$ and $y_0(t)$ is minimized [16]. Under this consideration, said parameter estimation problem may be described as follows.

Problem B2.6 (Parameter Estimation for FM Synthesizers)
Minimize:

$$J_6(\mathbf{x}) = \sum_{t=0}^{100} (y(\mathbf{x}, t) - y_0(t))^2$$

$$y(\mathbf{x}, t) = x_1 \sin(x_2\theta t) + x_2 \sin(x_4\theta t) + x_5 \sin(x_6\theta t)$$

Subject to:

$$-6.4 \le x_i \le 6.35, \quad i = 1, 2, 3, 4, 5, 6, ldots, 1x$$

am Design Problem describe each of the constrained evice [REF]. The previously mentioned real problems with those

$$0.1 \le x_i \le 10, \quad i = 2, 3$$

$$\theta = \frac{2\pi}{100}$$

B2.7 Optimal Capacitor Placement for the IEEE's 69-Bus Radial Distribution Networks

The Optimal Capacitor Placement (OCP) problem [17, 18] is a complex combinatorial optimization formulation in which the objective is to determine the number, location, type, and size of shunt capacitors that are to be placed in a Radial Distribution Network (RDN). In the problem, the main objective function involves the minimization of the network's total operation cost in terms of a certain amount of reactive compensation of the RDN [17].

In this problem, an RDN of 69 buses (which implies $n = 69$ dimensions) is assumed as it is illustrated in Fig. 2.9. Under this architecture, the model can be characterized as follows.

Problem B.7 (optimal capacitor placement for radial distribution networks)
Minimize:

$$J_7(\mathbf{x}) = K_{pt} P_{Loss} + K_{ic} N_q + \sum_{j=1}^{n} x_j k_j$$

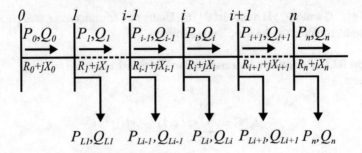

Fig. 2.9 Optimal capacitor placement design variables

Subject to:

$$x_j \in \{0, 150, 360, \ldots, 3900, 4050\}$$

$$\sum_{j=1}^{n_c} x_j \leq \sum_{i=1}^{n} x_i$$

In the model, K_p represents the equivalent cost per unit of power loss (\$/kW) and *PLoss* represents the total power loss of the distribution system (kW). *Qj* corresponds to the size of the shunt capacitor (kVAR) installed at bus "*j*". k_j symbolizes the cost per unit of reactive power (\$/kVAR). *t* is the operation time in hours, *Kic* is the installation cost (\$), and N_q is the number of compensation capacitors that are required to be installed in the distribution network. In the formulation, it is considered that $\mathbf{x} = [x_1 = Q_1, x_2 = Q_2, \ldots, xn = Q_n]$ and n_c is the number of buses required to be compensated.

References

1. Kennedy, J., Eberhart, R.: Particle swarm optimization. In: Proceedings of the IEEE International Conference on Neural Networks (ICNN'95), vol. 4, pp. 1942–1948. IEEE (1995)
2. Karaboga, D.: An Idea Based on Honey Bee Swarm For Numerical Optimization. Technical Report-TR06 (2005)
3. Gandomi, A.H., Yang, X.S., Alavi, A.H.: Cuckoo search algorithm: a metaheuristic approach to solve structural optimization problems. Eng. Comput. **29**(1), 17–35 (2013)
4. Yang, A.S.: Firefly algorithms for multimodal optimization. In: Stochastic Algorithms: Foundations and Applications, vol. 5792. Lecture Notes in Computer Science, pp. 169–178. Springer (20090
5. Cuevas, E., González, A., Zaldívar, D., Pérez-Cisneros, M.: An optimisation algorithm based on the behavior of locust swarms. Int. J. Bio-Inspired Comput. **7**(6), 402–407 (2015)
6. Cuevas, E., González, A., Fausto, F., Zaldívar, D., Pérez- Cisneros, M.: Multithreshold Segmentation by Using an Algorithm Based on the Behavior of Locust Swarms. Mathe. Problems Eng. vol. 2015, Article ID 805357, 25 pages (2015)
7. Cuevas, E., Gálvez, J., Avalos, O.: Parameter estimation for chaotic fractional systems by using the locust search algorithm. Computación y Sistemas **21**(2) (2017)

8. González, A., Cuevas, E., Fausto, F., Valdivia, A., Rojas, R.: A template matching approach based on the behavior of swarms of locust. Appl. Intell. **47**(4), 1087–1098 (2017)
9. Yang, X.-S.: A new metaheuristic bat-inspired algorithm,. In: Gonzalez, J.R., Pelta, D.A., Cruz, C., Terrazas, G., Krasnogor, N. (eds) Nature Inspired Cooperative Strategies for Optimization (NICSO 2010), vol. 284 *of Studies in Computational Intelligence*, pp. 65–74. Springer, Berlin, Germany (2010)
10. Storn, R., Price, K.: Differential evolution—a simple and efficient adaptive scheme for global optimization over continuous spaces (1995)
11. Geem, Z.W., Kim, J.H., Loganathan, G.V.: A new heuristic optimization algorithm: harmony search. Simulation **76**(2), 60–68 (2001)
12. Sandgren, E.: Nonlinear integer and discrete programming in mechanical design optimization. J. Mech. Design **112**(2), 223 (1990)
13. Arora, J.S.: Introduction to Optimum Design. Academic Press, New York, NY, USA (1989)
14. Rao, S.S.: Engineering Optimization: Theory and Practice. Wiley (2009)
15. Koski, J.: Defectiveness of weighting method in multicriterion optimization of structures. Commun. Numerical Methods Eng. **1**(6), 333–337 (1985)
16. Das, S., Suganthan, P.N.: Problem Definitions and Evaluation Criteria for CEC 2011Competition on Testing Evolutionary Algorithms on Real World Optimization Problems (2010)
17. Flaih, F.M.F., Lin, X., Dawoud, S.M., Mohammed, M.A.: Distribution system reconfiguration for power loss minimization and voltage profile improvement using Modified particle swarm optimization. In: Proceedings of the 2016 IEEE PES Asia Pacific Power and Energy Engineering Conference, (APPEEC'16), pp. 120–124 (2016)
18. Díaz, P., Pérez-Cisneros, M., Cuevas, E., Camarena, O., Fausto, F., Gonzalez, A.: A swarm approach for improving voltage profiles and reduce power loss on electrical distribution networks. IEEE Access **6**, 1 (2018)
19. Topaz, C.M., Bernoff, A.J., Logan, S., Toolson, W.: A model for rolling swarms of locusts. Euro. Phys. J. Special Topics **157**(1), 93–109 (2008)
20. Topaz, C.M., D' Orsogna, M.R., Edelstein-Keshet, L., Bernoff, A.L.: Locust dynamics: behavioral phase change and swarming. PLoS Computat. Biol. **8**(8), Article ID e1002642 (2012)
21. Stower, W.: Photographic techniques for the analysis of locust "hopper" behavior. Anim. Behav. **11**(1), 198–205 (1963)
22. Buhl, J., Sumpter, D.J.T., Couzin, I.D., et al.: From disorder to order in marching locusts. Science **312**(5778), 1402–1406 (2006)
23. Ariel, G., Ophir, Y., Levi, S., Ben-Jacob, E., Ayali, A.: Individual pause-and-go motion is instrumental to the formation and maintenance of swarms of marching locust nymphs. PLoS ONE **9**(7) (2014)
24. Ariel, G., Ayali, A., Adler, F.R.: Locust collective motion and its modeling. PLoS Computat. Biol. **11**(12), e1004522 (2015)
25. Goldberg, D.E.: Genetic Algorithms in Search, Optimization, and Machine Learning (1989)
26. López Ibáñez, M., Dubois-Lacoste, J., Pérez Cáceres, L., Birattari, M., Stützle, T.: The irace package: iterated racing for automatic algorithm configuration. Operat. Res. Perspect. **3**, 43–58 (2016)
27. https://abc.erciyes.edu.tr/
28. Gehan, E.A.: A generalized Wilcoxon test for comparing arbitrarily singly-censored samples. Biometrika **52**, 203–223 (1965)
29. Díaz Cortés, M.-A., Cuevas, E., Rojas, R.: Intelligent Systems Reference Library 129 Engineering Applications of Soft Computing. Springer Verlag (2017)
30. Cuevas, E.: Block-matching algorithm based on harmony search optimization for motion estimation, Appl. Intell. **39**(1), 165–183 (2913)

31. Díaz-Cortés, M.-A., Ortega-Sánchez, N., Hinojosa, S., Cuevas, E., Rojas, R., Demin, A.: A multi-level thresholding method for breast thermograms analysis using Dragonfly algorithm. Infrared Phys. Technol. **93**, 346–361 (2018)
32. Díaz, P., Pérez-Cisneros, M., Cuevas, E., Hinojosa, S., Zaldivar, D.: An im-proved crow search algorithm applied to energy problems. Energies **11**(3), 571 (2018)

Chapter 3
A Metaheuristic Methodology Based on Fuzzy Logic Principles

Abstract Various methods are so complex to be handled quantitatively; However, human beings have achieved by using simple rules that are extracted from their experiences. Fuzzy logic resembles human reasoning in its use of information to generate inaccurate decisions. Diffuse logic incorporates an alternative way of processing that allows complex systems to be modeled using a high level of abstraction originating from human knowledge and experiences. Recently, several of the new evolutionary computing algorithms have been proposed with exciting results. Several of them use operators based on metaphors of natural or social elements that evolve candidate solutions. Although humans have demonstrated their potential to solve complicated optimization problems of everyday life, they are not mechanisms to include such aptitudes into an evolutionary optimization algorithm. In this chapter, a methodology to implement human intelligence based on strategy optimization is presented. Under this approach, a procedure carried out is codified in rules based on *Takagi-Sugeno* diffuse inference system. So, to implement fuzzy practices, they express the conditions under which candidate solutions are evolved into new positions. To show the capability and robustness of the proposed approach, it is compared to other well-known evolutionary methods. The comparison examines several benchmark functions (benchmark) that are generally considered within the literature of evolutionary algorithms. The results confirm a high performance of the method in the search for a global optimum of different benchmark functions.

3.1 Introduction

There are processes that humans can do much better than deterministic systems or computers, such as obstacle avoidance while driving or planning a strategy. This may be due to our unique reasoning capabilities and complex cognitive processing. Although processes can be complex, humans undertake them by using simple rules of thumb extracted from their experiences. Fuzzy logic [1] is a practical alternative for a variety of challenging applications since it provides a convenient method for constructing systems via the use of heuristic information. The heuristic information

© Springer Nature Switzerland AG 2021
E. Cuevas et al., *Metaheuristic Computation: A Performance Perspective*,
Intelligent Systems Reference Library 195,
https://doi.org/10.1007/978-3-030-58100-8_3

may come from a system-operator who has directly interacted with the process. In the fuzzy logic design methodology, this operator is asked to write down a set of rules on how to manipulate the process. We then incorporate these into a fuzzy system that emulates the decision-making process of the operator [2]. For this reason, the partitioning of the system behavior into regions is an important characteristic of a fuzzy system [3]. In each region, the characteristics of the system can be simply modeled using a rule that associates the region under which certain actions are performed [4]. Typically, a fuzzy model consists of a rule base, where the information available is transparent and easily readable. The fuzzy modeling methodology has been largely exploited in several fields such as pattern recognition [5, 6], control [7, 8] and image processing [9, 10]. Recently, several optimization algorithms based on random principles have been proposed with interesting results. Such approaches are inspired by our scientific understanding of bio-logical or social systems, which at some abstraction level can be represented as optimization processes [11]. These methods mimic the social behavior of bird flocking and fish schooling in the Particle Swarm Optimization (PSO) method [12], the cooperative behavior of bee colonies in the Artificial Bee Colony (ABC) technique [13], the improvisation process that occurs when a musician searches for a better state of harmony in the Harmony Search (HS) [14], the attributes of bat behavior in the Bat Algorithm (BAT) method [15], the mating behavior of firefly insects in the Firefly (FF) method [16], the social behaviors of spiders in the Social Spider Optimization (SSO) [17], the characteristics of animal behavior in a group in the Collective Animal Behavior (CAB) [18] and the emulation of the differential and conventional evolution in species in the Differential Evolution (DE) [19] and Genetic Algorithms (GA) [20] respectively. On the other hand, the combination of fuzzy systems with meta-heuristic algorithms has recently attracted the attention in the Computational Intelligence community. As a result of this integration, a new class of systems known as Evolutionary Fuzzy Systems (EFSs) [21, 22] has emerged. These approaches basically consider the automatic generation and tuning of fuzzy systems through a learning process based on a metaheuristic method. The EFSs approaches reported in the literature can be divided into two classes [21, 22]: tuning and learning. In a tuning approach, a metaheuristic algorithm is applied to modify the parameters of an existent fuzzy system, without changing its rule base. Some examples of tuning in EFSs include the calibration of fuzzy controllers [23, 24], the adaptation of type-2fuzzy models [25] and the improvement of accuracy in fuzzy models [26, 27]. In learning, the rule base of a fuzzy system is generated by a metaheuristic algorithm, so that the final fuzzy system has the capacity to accurately reproduce the modeled system. There are several examples of learning in EFSs, which consider different types of problems such as the selection of fuzzy rules with member-ship functions [28, 29], rule generation [30, 31] and determination of the entire fuzzy structure [32–34]. The proposed method cannot be considered a EFSs approach, since the fuzzy system, used as optimizer, is not automatically generated or tuned by a learning procedure. On the contrary, its design is based on expert observations extracted from the optimization process. Therefore, the number of rules and its configuration are fixed, remaining static during its operation. Moreover, in a typical EFSs scheme, a metaheuristic algorithm is used to find an optimal base rule for a fuzzy system with regard to an evaluation function. Different to such approaches, in our method, a fuzzy system is employed to obtain the optimum value of an optimization problem. Hence, the produced Fuzzy system directly acts as any other metaheuristic algorithm conducting the optimization strategy implemented in its rules. A metaheuristic algorithm is conceived as a high-level problem-independent methodology that consists of a set of guidelines and operations to develop an optimization strategy. In this chapter, we describe how the fuzzy logic design methodology can be used to construct algorithms for optimization tasks. As opposed to "conventional" metaheuristic approaches where the focus is on the design of optimization operators that emulate a natural or social process, in our approach we focus on gaining an intuitive under-standing of how to conduct an efficient search strategy to model it directly into a fuzzy system. Although sometimes unnoticed, it is well understood that human heuristics play an important role in optimization methods. It must be acknowledged that metaheuristic approaches use human heuristics to tune their corresponding parameters or to select the appropriate algorithm for a certain problem [35]. Under such circumstances, it is important to ask the following questions: How much of the success may be assigned to the use of a certain meta-heuristic approach? How much should be attributed to its clever heuristic tuning or selection? Also, if we exploit the use of human heuristic information throughout the entire design process, can we obtain higher performance optimization algorithms? The use of fuzzy logic for the construction of optimization methods presents several advantages. (A) Generation. "Conventional" metaheuristic approaches reproduce complex natural or social phenomena. Such a reproduction involves the numerical modeling of partially-known behaviors and non-characterized operations, which are sometimes even unknown [36]. Therefore, it is notably complicated to correctly model even very simple metaphors. On the other hand, fuzzy logic provides a simple and well-known method for constructing systems via the use of human knowledge [37]. (B) Transparency. The metaphors used by metaheuristic approaches lead to algorithms that are difficult to understand from an optimization perspective. Therefore, the metaphor can-not be directly interpreted as a consistent search strategy [36]. On the other hand, fuzzy logic generates fully interpretable models whose content expresses the search strategy as humans can conduct it [38]. (C) Improvement. Once designed, metaheuristic methods maintain the same procedure to produce candidate solutions. Incorporating changes to improve the quality of candidate solutions is very complicated and severely damages the conception of the original metaphor [36]. As human experts interact with an optimization process, they obtain a better understanding of the correct search strategies that allow finding the optimal solution. As a result, new rules are obtained so that their inclusion in the existing rule base improves the quality of the original search strategy. Under the fuzzy logic methodology, new rules can be easily incorporated to an already existent system. The addition of such rules allows the capacities of the original system to be extended [39]. In this chapter, a methodology to implement human-knowledge-based optimization strategies is presented. In the scheme, a Takagi-Sugeno Fuzzy inference system [40] is used to reproduce a specific search strategy generated by a human expert. Therefore, the number of rules and its configuration only depend on the expert experience without considering any learning rule process. Under these conditions, each fuzzy rule represents an expert observation that models the conditions under which candidate solutions are modified in order to reach the optimal location. To exhibit the performance and robustness of the proposed method, a comparison to other well-known optimization methods is conducted. The comparison considers several standard benchmark functions which are typically found in the literature of metaheuristic optimization. The results suggest a high performance of the proposed methodology in comparison to existing optimization strategies 2.

3.2   Fuzzy Logic and Reasoning Models

This section presents an introduction to the main fuzzy logic concepts. The discussion particularly considers the *Takagi-Sugeno* Fuzzy inference model [40].

3.2.1   Fuzzy Logic Concepts

A fuzzy set (A) [1] is a generalization of a Crisp or Boolean set, which is defined in a universe of discourse X. A is a linguistic label which defines the fuzzy set through the word A. Such a word defines show a human expert perceives the variable X in relationship to A. The fuzzy set (A) is characterized by a membership function $\mu_A(x)$ which provides a measure of degree of similarity of an element x from X to the fuzzy set A. It takes values in the interval [0,1], that is:

$$\mu_A(x) : x \rightarrow [0, 1] \tag{3.1}$$

Therefore, a generic variable x_c can be represented using multiple fuzzy sets $\{A_1^c, A_2^c, \ldots, A_m^c\}$, each one modeled by a membership function $\{\mu A_1^c(x_c), \mu A_2^c(x_c), \ldots, \mu A_m^c(x_c)\}$.

A fuzzy system is a computing model based on the concepts of fuzzy logic. It includes three conceptual elements: a rule base, which contains a selection of fuzzy rules; a database, which defines the membership functions used by the fuzzy rules; and a reasoning mechanism, which performs the inference procedure. There are two different inference fuzzy systems: Mamdani [41] and *Takagi-Sugeno* (TS) [40]. The central difference between the two inference models is in the consequent section of the fuzzy systems. In the Mamdani model, all of the structure of the fuzzy system has linguistic variables and fuzzy sets. However, the consequent section of the TS model consists of mathematical functions. Different to the Mamdani structure, the TS model provides computational efficiency and mathematical simplicity in the rules [42]. Therefore, in order to obtain higher modelling accuracy with fewer rules, the TS fuzzy model is a good candidate that obtains better models when the rules are described as functional associations defined in several local behaviors [42, 43]. Since the available knowledge for the design of the fuzzy system conceived in our approach includes functional, local behaviors, the TS inference model has been used in this work for the system modeling.

3.2.2 The Takagi-Sugeno (TS) Fuzzy Model

TS fuzzy systems allow us to describe complicated nonlinear systems by decomposing the input space into several local behaviors, each of which is represented by a simple regression model [3]. The main component of a TS fuzzy system is the set of its K fuzzy rules. They code the human knowledge that explains the performance of the actual process. Each rule denoted by R^i relates the input variables to a consequence of its occurrence. A typical TS fuzzy rule is divided in two parts: Antecedent (I) and consequent (II), which are described as follows:

$$R^i : \overbrace{\text{IF } x_1 \text{ is } A_p^1 \text{ and } x_2 \text{ is } A_p^2, \ldots, \text{ and } x_n \text{ is } A_r^n \text{ Then}}^{\text{I}} \underbrace{y_i = g_i(\mathbf{x})}_{\text{II}} \quad (3.2)$$

$$i = 1, 2, \ldots, K$$

where $\mathbf{x} = [x_1, x_2, \ldots, x_n]^T$ is the n-dimensional input variable and y_i represents the output rule. $g(\mathbf{x})$ is a function which can be modeledby any function as long as it can appropriately describe the behavior of the system within the fuzzy region specified by the antecedent of rule i. In Eq. (3.2), p, q and r symbolize one fuzzy set which models the behavior of variables x_1, x_2 and x_n, respectively.

3.2.2.1 Antecedent (I)

The antecedent is a logical combination of simple prepositions of the form "x_e is A_d^e". Such a preposition, modeled by the membership function $\mu_{A_d^e}(x_e)$, provides a measure of degree of similarity between x_e and the fuzzy set A_d^e. Since the antecedent is concatenated by using the "and" connector, the degree of fulfilment of the antecedent $\beta_i(\mathbf{x})$ is calculated using a t-norm operator such as the minimum:

$$\beta_i(\mathbf{x}) = min\left(\mu_{A_p^1}(x_1), \mu_{A_p^2}(x_2), \ldots, \mu_{A_r^n}(x_n)\right) \quad (3.3)$$

3.2.2.2   Consequent (II)

$g_i(\mathbf{x})$ is a function which can be modeled by any function as long as it can appropriately describe the behavior of the system with in the fuzzy region specified by the antecedent of rule i.

3.2.2.3   Inference in the TS Model

The global output y of a TS fuzzy system is composed as the con-catenation of the local behaviors, and can be seen as the weighted mean of the consequents:

$$y = \frac{\sum_{i=1}^{K} \beta_i(\mathbf{x}) \cdot y_i}{\sum_{i=1}^{K} \beta_i(x)} \tag{3.4}$$

where $\beta_i(\mathbf{x})$ is the degree of fulfillment of the ith rule's antecedentand yiis the output of the consequent model of that rule. Figure 3.1 shows the fuzzy reasoning procedure for a TS fuzzy system with two rules. The example considers two variables (x_1, x_2) and only two membership functions (I and II) for each variable. Now, it should be clear that the spirit of fuzzy logic systems resembles that of "divide and conquer". Therefore, the antecedent of a fuzzy rule defines a local fuzzy region, while the consequent describes the behavior within the region.

3.3   The Proposed Methodology

Since there is no specific solution for several kinds of complex problems, human experts often follow a trial-and-error approach to solve them. Under this process, humans obtain experience as the knowledge gained through the interaction with the problem. In general, a fuzzy system is a model that emulates the decisions and behavior of a human that has specialized knowledge and experience in a particular field. Therefore, a fuzzy system is then presumed to be capable of reproducing

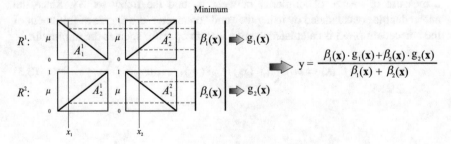

Fig. 3.1 TS Fuzzy model

the behavior of a target system. For example, if the target system is a human operator in charge of a chemical reaction process, then the fuzzy system becomes a fuzzy controller that can regulate the chemical process. Similarly, if the target system is a person who is familiar with optimization strategies and decision-making processes, then the fuzzy inference becomes a fuzzy expert system that can find the optimal solution to a certain optimization problem, as if the search strategy were conducted by the human expert. In this chapter, we propose a methodology for emulating human search strategies in an algorithmic structure. In this section, the fuzzy optimization approach is explained in detail. First, each component of the fuzzy system is described; then, the complete computational procedure is presented. Under a given set of circumstances, an expert provides a description of how to conduct an optimization strategy for finding the optimal solution to a generic problem using natural language. Then, the objective is to take this "linguistic" description and model it into a fuzzy system. The linguistic representation given by the expert is divided into two parts: (A) linguistic variables and (B) rule base formulation. (A) Linguistic variables describe the way in which a human expert perceives the circumstances of a certain variable in terms of its relative values. One example is the velocity that could be identified as low, moderate and high. (B) Rule base formulation captures the construction process of a set of IF-THEN associations. Each association (rule) expresses the conditions under which certain actions are performed. Typically, a fuzzy model consists of a rule base that maps fuzzy regions to actions. In this context, the contribution of each rule to the behavior of the fuzzy system will be different depending on the operating region.

Rules
R^1 : IF x_1 is A_1^1 and x_2 is A_2^2 Then $y_1 = g_1(\mathbf{x})$
R^2 : IF x_1 is A_2^1 and x_2 is A_1^2 Then $y_2 = g_2(\mathbf{x})$

3.3.1 Optimization Strategy

Most of the optimization methods have been designed to solve the problem of finding a global solution to a nonlinear optimization problem with box constraints in the following form [44]:

$$maximize \ f(x), \mathbf{x} = (x_1, \ldots, x_n) \in \mathbb{R}^n \qquad (3.5)$$

$$subjet \ to \ \mathbf{x} \in \mathbf{X}$$

where $f : \mathbb{R}^n \rightarrow \mathbb{R}$ is a nonlinear function $\{\mathbf{x} \in \mathbb{R}^n | l_i \leq x_i \leq u_i, i = 1, \ldots n\}$ is a bounded feasible search space, constrained by the lower (l_i) and upper (u_i)

limits. To solve the optimization problem presented in Eq. (3.5), from a population-based perspective [45], a set $\mathbf{P}^k\left(\{\mathbf{p}_1^k, \mathbf{p}_2^k, \ldots, \mathbf{p}_N^k\}\right)$ of N candidate solutions (individuals) evolves from an initial state ($k = 0$) to a maximum number of generations ($k = Maxgen$). In thefirst step, the algorithm initiates producing the set of N candidate solutions with values that are uniformly distributed between the pre-specified lower (l_i) and upper (u_i) limits. In each generation, agroup of evolutionary operations are applied over the population $\mathbf{P}^k$ to generate the new population $\mathbf{P}^{k+1}$. In the population, an individual $\mathbf{p}_1^k (i \in [1, \ldots, N])$ corresponds to a n-dimensional vector $\left\{p_{i,1}^k, p_{i,2}^k, \ldots, p_{i,n}^k\right\}$ where the dimensions represent the decision variables of the optimization problem to be solved. The quality of a candidate solution $\mathbf{p}_i^k$ is measured through an objective function $f\left(\mathbf{p}_i^k\right)$ whose value corresponds to the fitness value of $\mathbf{p}_i^k$. As the optimization process evolves, the best individual $\mathbf{g}(g_1, g_2, \ldots g_n)$ seen so-far is conserved, since it represents the current best avail-able solution. In the proposed approach, an optimization human-strategy is modelled in the rule base of a TS Fuzzy inference system, so that the implemented fuzzy rules express the conditions under which candidate solutions from $\mathbf{P}^k$ are evolved to new positions $\mathbf{P}^{k+1}$.

3.3.1.1 Linguistic Variables Characterization (A)

To design a fuzzy system from expert knowledge, it is necessary the characterization of the linguistic variables and the definition of a rule base. A linguistic variable is modeled through the use of membership functions. They represent functions which assign a numerical value to a subjective perception of the variable. The number and the shape of the membership functions that model a certain linguistic variable depend on the application context [46]. Therefore, in order to maintain the design of the fuzzy system as simple as possible, we characterize each linguistic variable by using only two membership functions [47]. One example is the variable velocity V that could be defined by the membership functions: low (μ_L) and high (μ_H). Such membership function is mutually exclusive or disjoint. Therefore, if $\mu_L = 0.7$, then $\mu_L = 0.3$. Assuming that the linguistic variable velocity "V" has a numerical value inside the interval from 0 to 100 revolutions per minute (rpm), μ_L and μ_H are characterized according to the membership functions shown in Fig. 3.2.

Fig. 3.2 Example of membership functions that characterize a linguistic variable

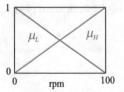

3.3.1.2 Rule Base Formulation (B)

Several optimization strategies can be formulated by using human knowledge. In this section, a simple search strategy is formulated considering basic observations of the optimization process. Therefore, the simplest search strategy is to move candidate solutions to search regions of the space where it is expected to find the optimal solution. Since the values of the objective function are only known in the positions determined by the candidate solutions, the locations with the highest probabilities of representing potential solutions are those located near the best candidate solution in terms of its fitness value. Taking this into consideration, a simple search strategy could be formulated by the following four rules:

1. **IF** the distance from $\mathbf{p}_i^k$ to $\mathbf{g}$ is short **AND** $f\left(\mathbf{p}_i^k\right)$ is good **THEN** $\mathbf{p}_i^k$ is moved towards (Attraction) $\mathbf{g}$. This rule represents the situation where the candidate solution $\mathbf{p}_i^k$ is moved to the best candidate solution seen so-far $\mathbf{g}$ in order to improve its fitness quality. Since the fitness values of $\mathbf{p}_i^k$ and g are good in comparison to other members of $\mathbf{P}^k$, the region between $\mathbf{p}_i^k$ and $\mathbf{g}$ maintains promising solutions that could improve $\mathbf{g}$. Therefore, with this movement, it is expected to explore the unknown region between $\mathbf{p}_i^k$ and $\mathbf{g}$. In order to show how each rule performs. Figure 3.3 shows a simple example which expresses the conditions under which action rules are executed. In the example, a population $\mathbf{P}^k$ of five candidate solutions is considered (see Fig. 3.3a). In the case of rule 1, as it is exhibited in Fig. 3.3b, the candidate solution $\mathbf{p}_5^k$ that fulfills the rule requirements is attracted to $\mathbf{g}$.

2. **IF** the distance from $\mathbf{p}_i^k$ to $\mathbf{g}$ is short **AND** $f\left(\mathbf{p}_i^k\right)$ is bad **THEN** $\mathbf{p}_i^k$ is moved away from (Repulsion) $\mathbf{g}$. In this rule, although the distance between $\mathbf{p}_i^k$ and $\mathbf{g}$ is short, the evidence shows that there are no good solutions between them. Therefore, the improvement of $\mathbf{p}_i^k$ is searched in the opposite direction of $\mathbf{g}$. A visual example of this behavior is presented in Fig. 3.3c.

3. **IF** the distance from $\mathbf{p}_i^k$ to $\mathbf{g}$ is large **AND** $f\left(\mathbf{p}_i^k\right)$ is good **THEN** $\mathbf{p}_i^k$ is refined. Under this rule, a good candidate solution $\mathbf{p}_i^k$ that is far from $\mathbf{g}$ is refined by searching within its neighborhood. The idea is to improve the quality of competitive candidate solutions which have already been found (exploitation). Such a scenario is presented in Fig. 3.3d where the original candidate solution $\mathbf{p}_2^k$ is substituted by a new position $\mathbf{p}_2^{k+1}$ which is randomly produced within the neighborhood of $\mathbf{p}_2^k$.

4. **IF** the distance from $\mathbf{p}_i^k$ to $\mathbf{g}$ is large **AND** $f\left(\mathbf{p}_i^k\right)$ is bad **THEN** a new position is randomly chosen. This rule represents the situation in Fig. 3.3e where the candidate solution $\mathbf{p}_4^k$ is so bad and so far from g that is better to replace it by other solution $\left(\mathbf{p}_4^{k+1}\right)$ randomly produced within the search space $\mathbf{X}$.

Each of the four rules listed above is a "linguistic rule" which contains only linguistic information. Since linguistic expressions are not well-defined descriptions of the values that they represent, linguistic rules are not accurate. They represent only conceptual ideas about how to achieve a good optimization strategy according to the

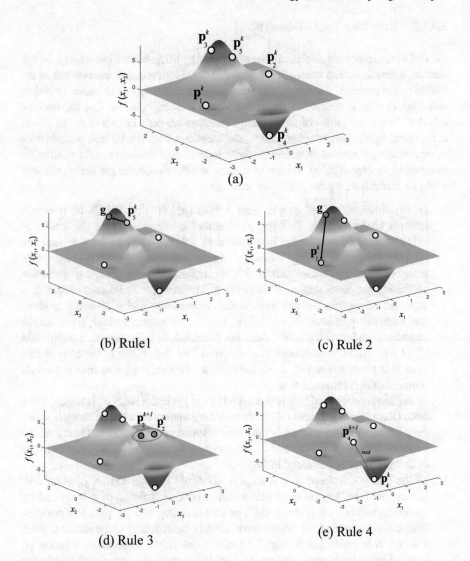

Fig. 3.3 Visual example that expresses the conditions under which action rules are executed. **a** Current configuration of the candidate solution population $\mathbf{P}^k$, **b** rule 1, **c** rule 2, **d** rule 3 and **e** rule 4

human perspective. Under such conditions, it is necessary to define the meaning of their linguistic descriptions from a computational point of view.

3.3.1.3 Implementation of the TS Fuzzy System

In this section, we will discuss the implementation of the expert knowledge concerning the optimization process in a TS fuzzy system.

(I) Membership functions and antecedents

In the rules, two different linguistic variables are considered, distance from de candidate solution $\mathbf{p}_i^k$ to the best solution $\mathbf{g}$ $\left(D\left(\mathbf{p}_i^k, \mathbf{g}\right)\right)$ and the fitness value of the candidate solution $f\left(\mathbf{p}_i^k\right)$. Therefore, $D\left(\mathbf{p}_i^k, \mathbf{g}\right)$ is characterized by two membership functions: short and large (see 3.3.1.1). On the other hand, $f\left(\mathbf{p}_i^k\right)$ is modeled by the membership functions good and bad. Figure 3.4 shows the fuzzy membership functions for both linguistic variables.

The distance $D\left(\mathbf{p}_i^k, \mathbf{g}\right)$ is defined as the Euclidian distance $\left\|\mathbf{g} - \mathbf{p}_i^k\right\|$. Therefore, as it is exhibited in Fig. 3.4a, two complementary membership functions define the relative distance $D\left(\mathbf{p}_i^k, \mathbf{g}\right)$: short (**S**) and large (**L**). Their support values are 0 and d_{max}, where d_{max} represents the maximum possible distance delimited by the search space **X** which is defined as follows:

$$d_{max} = \sqrt{\sum_{s=1}^{d} (u_s - l_s)^2} \tag{3.6}$$

where d represents the number of dimensions in the search space **X**. In the case of $f\left(\mathbf{p}_i^k\right)$, two different membership functions define its relative value: bad (**B**) and good (**G**). Their support values are f_{min} and f_{max}. These values represent the minimum and maximum fitness values seen so-far. Therefore, they can defined as following:

$$f_{min} = \min f\left(\mathbf{p}_i^k\right) \text{ and } i \in \{1, 2, \ldots, N\}, k \in \{1, 2, \ldots, gen\}$$
$$f_{max} = \max f\left(\mathbf{p}_i^k\right) \text{ and } i \in \{1, 2, \ldots, N\}, k \in \{1, 2, \ldots, gen\} \tag{3.7}$$

From Eq. (3.7), it is evident that $f_{max} = f(\mathbf{g})$. If a new minimum or maximum value of $f\left(\mathbf{p}_i^k\right)$ is detected during the evolution process, it replaces the past values of f_{min} or f_{max}. Figure 3.4b shows the membership functions that describe $f\left(\mathbf{p}_i^k\right)$.

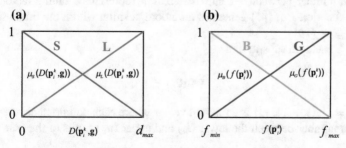

Fig. 3.4 Membership functions for **a** distance $D\left(\mathbf{p}_i^k, \mathbf{g}\right)$ and **b** for $f\left(\mathbf{p}_i^k\right)$

Table 3.1 Degree of fulfilment of the antecedent $\beta_w(\mathbf{x})$ for each rule ($w \in [1, 2, 3, 4]$)

Rule	Degree of fulfilment $\beta_w(\mathbf{x})$
1	$\beta_1(\mathbf{p}_i^k) = \min(\mu_S(D(\mathbf{p}_i^k, \mathbf{g})), \mu_G(f(\mathbf{p}_i^k)))$
2	$\beta_2(\mathbf{p}_i^k) = \min(\mu_S(D(\mathbf{p}_i^k, \mathbf{g})), \mu_B(f(\mathbf{p}_i^k)))$
3	$\beta_3(\mathbf{p}_i^k) = \min(\mu_L(D(\mathbf{p}_i^k, \mathbf{g})), \mu_G(f(\mathbf{p}_i^k)))$
4	$\beta_4(\mathbf{p}_i^k) = \min(\mu_L(D(\mathbf{p}_i^k, \mathbf{g})), \mu_B(f(\mathbf{p}_i^k)))$

Considering the membership functions defined in Fig. 3.4, the degree of fulfilment of the antecedent $\beta_w(\mathbf{x})$ for each rule ($w \in [1, 2, 3, 4]$) is defined in Table 3.1.

(II) Actions or consequents

Actions or Consequents are functions which can be modeled by any function as long as it can appropriately describe the desired behavior of the system within the fuzzy region specified by the antecedent of a rule i ($i \in [1, 2, 3, 4]$). The consequents of the four rules are modeled by using the following behaviors.

Rule 1. Attraction

$$At(\mathbf{p}_i^k) = |f_{max} - f(\mathbf{p}_i^k)| \cdot (\mathbf{g} - \mathbf{p}_i^k) \cdot \alpha_1 \qquad (3.8)$$

where α_1 represents a tuning factor. Under this rule, the function $At(\mathbf{p}_i^k)$ produces a change of position in the direction of the attraction vector $(\mathbf{g} - \mathbf{p}_i^k)$. The magnitude depends on the difference of the fitness values between $\mathbf{g}$ and $\mathbf{p}_i^k$.

Rule 2. Repulsion

$$Rep(\mathbf{p}_i^k) = |f_{max} - f(\mathbf{p}_i^k)| \cdot (\mathbf{g} - \mathbf{p}_i^k) \cdot \alpha_2 \qquad (3.9)$$

Where α_2 represents a tuning factor.

Rule 3. Refining or perturbation.

$$Ref(\mathbf{p}_i^k) = |f_{max} - f(\mathbf{p}_i^k)| \cdot \mathbf{v} \cdot \gamma \qquad (3.10)$$

where $\mathbf{v} = \{v_1, v_2, \ldots, v_d\}$ is a random vector where each component represents a random number between -1 and 1 whereas γ represents a tuning factor. In this rule, the function $Ref(\mathbf{p}_i^k)$ generates a random position within the limits specified by $\pm |f_{max} - f(\mathbf{p}_i^k)|$.

Rule 4. Random substitution

$$Ran(\mathbf{p}_i^k) = \mathbf{r} \qquad (3.11)$$

where $\mathbf{r} = \{r_1, r_2, \ldots, r_d\}$ is a random vector where each compo-nent ru represents a random number between the lower (l_u) and upper (u_u) limits of the search space $\mathbf{X}$.

(III) *Inference of the TS model.*

The global change of position $\Delta \mathbf{p}_i^k$ of the TS fuzzy system is composed as the concatenation of the local behaviors produced by the four rules, and can be seen as the weighted mean of the consequents:

$$\Delta \mathbf{p}_i^k = \frac{At\left(\mathbf{p}_i^k\right) \cdot \beta_1\left(\mathbf{p}_i^k\right) + \text{Rep}\left(\mathbf{p}_i^k\right) \cdot \beta_2\left(\mathbf{p}_i^k\right) + \text{Ref}\left(\mathbf{p}_i^k\right) \cdot \beta_3\left(\mathbf{p}_i^k\right) + \text{Ran}\left(\mathbf{p}_i^k\right) \cdot \beta_4\left(\mathbf{p}_i^k\right)}{\beta_1\left(\mathbf{p}_i^k\right) + \beta_2\left(\mathbf{p}_i^k\right) + \beta_3\left(\mathbf{p}_i^k\right) + \beta_4\left(\mathbf{p}_i^k\right)}$$

$$(3.12)$$

Once $\Delta \mathbf{p}_i^k$ has been calculated, the new position $\mathbf{p}_i^{k+1}$ is calculated as follows:

$$\mathbf{p}_i^{k+1} = \mathbf{p}_i^k + \Delta \mathbf{p}_i^k \tag{3.13}$$

3.3.2 Computational Procedure

The proposed algorithm is implemented as an iterative process in which several operations are executed. Such operations can be summarized in the form of pseudo-code in Algorithm 3.1. The proposed method uses as input information the number of candidate solutions (N), the maximum number of generations (*Maxgen*), and the tuning parameters μ_1, μ_2, γ. Similar to other metaheuristic algorithms, in the first step (line 2), the algorithm initiates producing the set of N candidate solutions with values that are uniformly distributed between the pre-specified lower and upper limits. These candidate solutions represent the first population $\mathbf{P}^0$. After initialization, the best element $\mathbf{g}$ in terms of its fitness value is selected (line 3). Then, for each particle $\mathbf{p}_i^k$ its distance to the best value g is calculated (line 6). With $D\left(\mathbf{p}_i^k, \mathbf{g}\right)$ and $f\left(\mathbf{p}_i^k\right)$, the search optimization strategy implemented in the fuzzy system is applied (lines 7–9). Under such circumstances, the antecedents (line 7) and consequents (line 8) are computed while the final displacement $\Delta \mathbf{p}_i^k$ is obtained as a result of the operation performed by the TS model (line 9). Afterwards, the new position $\mathbf{p}_i^{k+1}$ is updated (line 10). Once the new population $\mathbf{P}^{k+1}$ is obtained as a result of the iterative operation of lines 6–10, the best value $\mathbf{g}$ is updated (line 12). This cycle is repeated until the maximum number the iterations *Maxgen* has been reached.

Algorithm 3.1 Summarized operations of the proposed Fuzzy method

1	**Input:** N, *Maxgen*, $\alpha_1, \alpha_2, \gamma$, k=0
2	$\mathbf{P}^k \leftarrow$ **Initialize** (N);
3	$\mathbf{g} \leftarrow$ **Select Best Particle** ($\mathbf{P}^k$);
4	**while** $k <$ *Maxgen* **do**
5	**for** (i=1; $i > N$; i++)
6	$D(\mathbf{p}_i^k, \mathbf{g}) \leftarrow$ **Calculate the Distance the Best** ($\mathbf{p}_i^k, \mathbf{g}$);
7	$[\beta_1, \beta_2, \beta_3, \beta_4] \leftarrow$ **Evaluate Antecedents** $D(\mathbf{p}_i^k, \mathbf{g}), f(\mathbf{p}_i^k)$;
8	$[At, Rep, Ref, Ran] \leftarrow$ **Evaluate Consequents** $(\mathbf{p}_i^k, \mathbf{g}), f(\mathbf{p}_i^k)$;
9	$\Delta\mathbf{p}_i^k \leftarrow$ **Inference TS** ($\beta_1, \beta_2, \beta_3, \beta_4, At, Rep, Ref, Ran$);
10	$\mathbf{p}_i^{k+1} \leftarrow \mathbf{p}_i^k + \Delta\mathbf{p}_i^k$
11	end **for**
12	$\mathbf{g} \leftarrow$ Select Best Particle ($\mathbf{P}^{k+1}$);
13	$k \leftarrow k + 1$
14	end **while**
15	output: $\mathbf{g}$

3.4 Discussion About the Proposed Methodology

In this section, several important characteristics of the proposed algorithm are discussed. First, in Sect. 3.4.1, interesting operations of the optimization process are analyzed. Next, in Sect. 3.4.2 the modelling properties of the proposed approach are high-lighted.

3.4.1 Optimization Algorithm

A metaheuristic algorithm is conceived as a high-level problem independent methodology that consists of a set of guidelines and operations to develop an optimization strategy. In the proposed methodology, a fuzzy system is generated based on expert observations about the optimization process. The final fuzzy system then performs various fuzzy logic operations to produce a new candidate solution $\mathbf{p}_i^{k+1}$ from the current solution $\mathbf{p}_i^k$. During this process, the following operations are involved:

1. Determination of the degree of membership between the input data $\left(D(\mathbf{p}_i^k, \mathbf{g}), f(\mathbf{p}_i^k)\right)$ and the defined fuzzy sets ("short & large" or "good & bad").
2. Calculation of the degree of relevance for each rule based on the degree of fulfilment $\beta_w(\mathbf{x})$ for each rule ($w \in [1, 2, 3, 4]$) in the antecedent part of the rule.
3. Evaluation of the consequent of each rule: *At, Rep, Ref, Ran.*

4. Derivation of the new candidate solution $\mathbf{p}_i^{k+1}$ based on the weighted mean of the consequent functions, according to the TS model.

Under such circumstances, the generated fuzzy system is applied over all candidate solutions from $\mathbf{P}^k$ in order to produce the new population $\mathbf{p}_i^{k+1}$. This procedure is iteratively executed until a termination criterion has been reached.

3.4.2 Modeling Characteristics

One of the methods used to solve optimization problems are metaheuristic algorithms. They have demonstrated its powerful characteristics in comparison with classical optimization techniques [58–60]. Such algorithms have been developed by a combination of deterministic models and randomness, mimicking the behavior of biological or social systems. Most of the metaheuristic methods divide the individual behavior into several processes which show no coupling among them [11, 36].

In the proposed methodology, the produced fuzzy system models a complex optimization strategy. This modeling is accomplished by a number of fuzzy IF-THEN rules, each of which describes the local behavior of the model. In particular, the rules express the conditions under which new positions are explored. In order to calculate a new candidate solution $\mathbf{p}_i^{k+1}$, the consequent actions of all rules are aggregated. In this way, all the actions are presented in the computation of a certain solution $\mathbf{p}_i^{k+1}$, but with different influence levels. By coupling local behaviors, fuzzy systems are able to reproduce complex global behaviors. An interesting example of such modeling characteristics is rule 1 and rule 2. If these rules are individually analyzed, the attraction and repulsion movements conducted by the functions are completely deterministic. However, when all rules are considered, rule 3 and rule 4 add randomness to the final position of $\mathbf{p}_i^{k+1}$.

3.5 Experimental Study

An illustrative set of 19 functions has been used to examine the performance of our approach. These test functions represent the base functions from the latest competition on single objective optimization problems at CEC2015 [48]. Tables 3.17, 3.18 and 3.19 in Appendix A show the benchmark functions employed in our experiments. These functions are ordered into three different classes: Unimodal (Table 3.17), multimodal (Table 3.18) and Hybrid (Table 3.19) test functions. In the tables, n represents the dimension in which the function is operated, $f(\mathbf{x}^*)$ characterizes the optimal value of the function in the position $\mathbf{x}^*$ and S is the defined search space.

The main objective of this section is to present the performance of the proposed algorithm on numeric optimization problems. Moreover, the results of our method are compared with some popular optimization algorithms by using the complete set of benchmark functions. The results of the proposed algorithm are verified by a statistical analysis of the experimental data.

The experimental results are divided into two sub-sections. In the first section, the performance of the proposed algorithm is evaluated with regard to its tuning parameters. In the second section, the overall performance of the proposed method is compared to six popular optimization algorithms based on random principles.

3.5.1   Performance Evaluation with Regard to Its Own Tuning Parameters

The three parameters of the rules α_1, α_2 and γ affect the expected performance of the proposed fuzzy optimization algorithm. In this sub-section we analyze the behavior of the proposed algorithm considering the different settings of these parameters. All experiments have been executed on a Pentium dual-core computer with 2.53-GHz and 4-GB RAM under MATLAB® 8.3. For the sake of simplicity, only the functions from f_1 to f_{14} (unimodal and multimodal) have been considered in the tuning process. In the simulations, all the functions operate with a dimension $n = 30$. As an initial condition, the parameters α_1, α_2 and γ are set to their default values rules $\alpha_1 = 1.4$, $\alpha_2 = 0.05$ and $\gamma = 0.005$. Then, in our analysis, the three parameters are evaluated one at a time, while the other two parameters remain fixed to their default values. To minimize the stochastic effect of the algorithm, each benchmark function is executed independently a total of 10 times. As a termination criterion, the maximum number of iterations (*Maxgen*) is set to 1000. In all simulations, the population size N is fixed to 50 individuals.

In the first stage, the behavior of the proposed algorithm is analyzed considering different values for α_1. In the analysis, the values of α_1 vary from 0.6 to 1.6 whereas the values of α_2 and γ remainfixed at 0.05 and 0.005, respectively. In the simulation, the proposed method is executed independently 30 times for each value of α_1 on each benchmark function. The results are registered in Table 3.2. These values represent the average best fitness values $(\bar{f})$ and the standard deviations (σ_f) obtained in terms of a certain parameter combination of α_1, α_2 and γ. From Table 3.2, we can conclude that the proposed fuzzy algorithm with $\alpha_1 = 1.4$ maintains the best performance on functions $f_1 - f_9$, and f_{11}. Under this configuration, the algorithm obtains the best results in 9 out of 14 functions. On the other hand, when the parameter α_1 is set to any other value, the performance of the algorithm is inconsistent, producing generally bad results.

In the second stage, the performance of the proposed algorithm is evaluated considering different values for α_2. In the experiment,the values of α_2 are varied from 0.01 to 0.1 whereas the values of α_1 and γ remain fixed at 1.4 and 0.005, respectively. The statistical results$\alpha_2 v$ are presented in Table 3.3. From Table 3.3, it is clear that our fuzzy optimization algorithm with $\alpha_2 = 0.05$ outperforms the other parameter configurations. Under this configuration, the algorithm obtains the best

Table 3.2 Experimental results obtained by the proposed algorithm using different values of α_1

α_1		0.6	0.7	0.8	0.9	1	1.2	1.3	1.4	1.5	1.6
f_1	$\bar{f}$	6.95E−55	7.74E−89	3.97E−167	1.01E−39	1.02E−193	0.00E+00	4.26E−29	**3.08E−281**	1.15E−28	6.85E−28
	σ_f	3.67E−54	4.24E−88	0.00E+00	5.54E−39	0.00E+00	0.00E+00	2.19E−28	**0.00E+00**	3.41E−28	1.08E−27
f_2	$\bar{f}$	6.10E−23	1.14E−53	1.12E−124	2.49E−139	1.08E−158	7.16E−22	1.31E−78	**2.66E−207**	2.03E−15	7.52E+00
	σ_f	3.34E−22	6.07E−53	4.81E−124	1.36E−138	0.00E+00	0.00E+00	7.18E−78	**0.00E+00**	4.31E−15	2.35E+01
f_3	$\bar{f}$	4.69E−10	1.80E−17	2.22E−22	3.23E−22	2.00E−27	3.89E−21	4.11E−27	**1.00E−27**	2.68E−18	1.93E−11
	σ_f	1.62E−09	6.81E−17	8.87E−22	1.64E−21	4.73E−27	4.96E−24	9.21E−27	**1.50E−27**	1.12E−17	5.52E−11
f_4	$\bar{f}$	1.55E−23	1.48E−30	1.51E−130	4.64E−180	2.84E−112	2.66E−23	2.00E−183	**3.85E−220**	9.09E-16	5.16E−15
	σ_f	7.09E−23	8.12E−30	8.25E−130	0.00E+00	1.56E−111	6.13E+19	0.00E+00	**0.00E+00**	2.76E−15	6.91E−15
f_5	$\bar{f}$	2.85E+01	2.85E+01	2.85E+01	2.85E+01	2.85E+01	3.31E+18	2.85E+01	**3.04E−03**	2.85E+01	2.85E+01
	σ_f	4.38E−02	3.86E−02	3.87E−02	4.37E−02	3.04E−02	1.75E−02	4.50E−02	**3.02E−02**	4.58E−02	4.72E−02
f_6	$\bar{f}$	2.15E−02	1.05E−02	1.19E−02	1.57E−02	1.59E−02	1.67E−02	1.69E−02	**7.94E−03**	1.98E−02	1.92E−02
	σ_f	1.90E−02	3.86E−03	1.13E−02	1.56E−02	1.49E−02	1.37E−02	1.49E−02	**1.89E−03**	1.80E−02	9.92E−03
f_7	$\bar{f}$	8.98E−03	3.22E−03	2.22E−03	1.88E−03	1.75E−03	2.07E−03	1.79E−03	**1.36E−03**	1.59E−03	1.64E−03
	σ_f	1.27E−02	2.65E−03	2.17E−03	1.48E−03	1.62E−03	2.26E−03	1.92E−03	**1.10E−03**	1.45E−03	1.50E−03
f_8	$\bar{f}$	−4.95E+03	−5.12E+03	−5.01E+04	−5.14E+04	−4.96E+03	−5.02E+03	−5.23E+03	**−5.58E+04**	−4.94E+03	−5.21E+03
	σ_f	4.36E+02	5.09E+02	4.15E+02	5.55E+02	5.08E+02	4.78E+02	4.24E+02	**4.10E+02**	4.85E+02	4.60E+02
f_9	$\bar{f}$	6.20E+01	2.91E+01	1.70E+01	1.57E+01	5.94E+00	5.21E+00	5.86E+00	**4.76E−01**	9.94E+00	2.02E+01
	σ_f	6.08E+01	5.32E+01	4.17E+01	4.27E+01	3.17E+01	2.77E+01	2.90E+01	**2.38E+00**	3.77E+01	5.22E+01
f_{10}	$\bar{f}$	8.70E−15	**7.16E−15**	7.99E−15	9.18E−15	9.41E−15	1.04E−14	1.19E−14	8.47E−15	1.07E−14	1.38E−14
	σ_f	5.39E−15	3.92E−15	3.61E−15	**2.53E−15**	2.57E−15	5.22E−15	6.00E−15	3.82E−15	6.64E−15	7.83E−15
f_{11}	$\bar{f}$	**0.00E+00**	**0.00E+00**	**0.00E+00**	**0.00E+00**	**0.00E+00**	**0.00E+00**	**0.00E+00**	**0.00E+00**	**0.00E+00**	3.46E−05

(continued)

Table 3.2 (continued)

α_1		0.6	0.7	0.8	0.9	1	1.2	1.3	1.4	1.5	1.6
	σ_f	**0.00E+00**	**0.00E+00**	**0.00E+00**	**0.00E+00**	**0.00E+00**	**0.00E+00**	**0.00E+00**	**0.00E+00**	**0.00E+00**	1.32E−04
f_{12}	$\bar{f}$	8.76E−02	8.32E−02	8.42E−02	7.82E−02	**7.70E−02**	8.73E−02	9.45E−02	9.59E−01	3.84E+00	6.49E+02
	σ_f	2.74E−02	3.35E−02	4.73E−02	2.30E−02	**1.80E−02**	2.78E−02	2.15E−02	3.26E+00	8.79E+00	2.65E+03
f_{13}	$\bar{f}$	**2.88E−01**	3.30E−01	3.69E−01	3.77E−01	3.99E−01	3.72E−01	4.27E−01	3.91E−01	2.44E+01	1.83E+06
	σ_f	**1.42E−01**	1.03E−01	1.63E−01	1.57E−01	2.20E−01	1.16E−01	3.78E−01	1.76E−01	9.67E+01	1.00E+07
f_{14}	$\bar{f}$	−8.43E+02	−8.33E+02	−8.31E+02	−8.29E+02	−8.43E+02	−8.97E+02	**−8.98E+02**	−8.90E+02	−8.86E+02	−8.84E+02
	σ_f	1.14E−01	9.37E+00	**1.06E+01**	1.12E+01	1.68E+01	2.54E+01	2.19E+01	1.80E+01	2.13E+01	2.48E+01

Bold values correspond to the best value obtained for each function

Table 3.3 Experimental results obtained by the proposed algorithm using different values of α_2

α_2		0.01	0.02	0.03	0.04	0.05	0.06	0.07	0.08	0.09	0.1
f_1	$\bar{f}$	1.01E−39	3.25E−28	3.39E−28	2.35E−28	**5.18E−49**	1.48E−28	1.36E−28	2.61E−28	2.15E−28	2.29E−28
	σ_f	5354−39	9.09E−28	1.02E−27	7.44E−28	**2.34E−48**	5.19E−28	4.95E−28	8.03E−28	6.36E−28	6.94E−28
f_2	$\bar{f}$	6.25E−16	3.61E−16	1.67E−22	2.66E−20	7.16E−22	1.15E−16	7.85E−16	3.29E−17	4.82E−16	**6.57E−30**
	σ_f	2.85E−15	1.97E−15	9.11E−22	1.45E−19	3.89E−21	5.41E−19	2.97E−15	1.53E−16	2.47E−15	**3.59E−29**
f_3	$\bar{f}$	4.96E−24	9.65E−27	9.29E−26	2.22E−25	**1.97E−27**	2.52E−23	2.81E−21	4.94E−22	7.99E−23	6.69E−21
	σ_f	2.66E−23	3.49E−26	4.33E−25	1.12E−24	**2.02E−27**	1.37E−22	1.53E−20	2.64E−21	2.25E−22	2.13E−20
f_4	$\bar{f}$	1.96E−15	1.20E−15	3.44E−17	5.99E−18	**3.08E−29**	1.92E−20	4.91E−26	6.13E−19	3.98E−16	1.42E−28
	σ_f	5.12E−15	4.45E−15	1.33E−16	3.28E−17	**1.69E−28**	9.01E−20	2.69E−25	3.31E−18	2.18E−15	7.76E−28
f_5	$\bar{f}$	2.85E+01	2.85E+01	2.55E+01	2.85E+01	**1.99E−04**	2.85E+01	2.85E+01	2.85E+01	2.85E+01	2.85E+01
	σ_f	4.45E−02	4.42E−02	4.38E−02	4.42E−02	**3.74E−02**	5.02E−02	3.45E−02	4.12E−02	5.02E−02	4.04E−02
f_6	$\bar{f}$	1.55E−02	1.47E−02	2.11E−02	2.15E−02	**1.07E−02**	1.73E−02	1.94E−02	1.78E−02	2.35E−01	2.04E−02
	σ_f	9.87E−03	5.13E−03	2.10E−02	4.72E−03	**1.67E−02**	6.85E−03	**1.90E−02**	7.59E−03	1.17E+00	2.21E−02
f_7	$\bar{f}$	**1.03E−03**	1.56E−03	1.09E−03	1.42E−03	1.36E−03	2.17E−03	1.88E−03	2.12E−03	2.53E−03	2.55E−03
	σ_f	8.62E−04	1.60E−03	7.41E−04	**1.07E−03**	1.10E−03	1.22E−03	1.56E−03	2.12E−03	3.04E−03	2.10E−03
f_8	$\bar{f}$	−5.10E+03	−5.24E+03	−5.17E+03	−5.00E+03	−5.13E+03	−5.01E+03	**−5.29E+03**	−5.11E+03	−5.14E+03	−5.14E+03
	σ_f	5.08E+02	4.49E+02	4.65E+02	3.83E+02	4.77E+02	**3.68E+02**	5.35E+02	4.36E+02	5.03E+02	4.41E+02
f_9	$\bar{f}$	8.98E+00	3.41E−02	5.91E+00	**9.16E−02**	4.76E−01	8.28E+00	7.75E−02	4.07E+00	1.64E+01	5.83E+00
	σ_f	3.42E+01	**1.87E−01**	3.14E+01	2.81E−01	2.38E+00	3.15E+01	2.61E−01	2.18E+01	5.02E+01	3.13E+01
f_{10}	$\bar{f}$	1.12E−14	1.01E−14	1.10E−14	1.17E−14	**8.47E−15**	9.30E−15	9.89E−15	1.21E−14	1.21E−14	1.20E−14
	σ_f	4.88E−15	3.82E−15	4.86E−15	7.26E−15	**3.44E−15**	4.71E−15	3.82E−15	6.19E−15	7.11E−15	6.03E−15
f_{11}	$\bar{f}$	**0.00E+00**	**0.00E+00**	**0.00E+00**	**0.00E+00**	**0.00E+00**	**0.00E+00**	**0.00E+00**	**0.00E+00**	**0.00E+00**	**0.00E+00**

(continued)

Table 3.3 (continued)

α_2		0.01	0.02	0.03	0.04	0.05	0.06	0.07	0.08	0.09	0.1
	σ_f	**0.00E+00**	**0.00E+00**	**0.00E+00**	**0.00E+00**	**0.00E+00**	**0.00E+00**	**0.00E+00**	**0.00E+00**	**0.00E+00**	**0.00E+00**
f_{12}	$\bar{f}$	1.03E+00	7.77E−01	**9.55E−02**	1.93E+00	9.59E−01	4.14E−01	1.17E+00	1.26E+00	1.60E+00	6.41E+00
	σ_f	3.80E+00	2.63E+00	**3.50E−02**	4.30E+00	3.26E+00	1.75E+00	3.56E+00	4.41E+00	4.70E+00	2.28E+01
f_{13}	$\bar{f}$	3.54E−01	4.08E−01	1.69E+00	2.38E+00	**3.91E−01**	4.20E−01	1.17E+00	1.12E+00	8.94E−01	1.34E+00
	σ_f	1.91E−01	2.18E−01	5.25E+00	7.65E+00	**1.76E−01**	1.69E−01	4.03E+00	3.38E+00	2.53E+00	4.99E+00
f_{14}	$\bar{f}$	−8.85E+02	−8.84E+02	−8.91E+02	−8.88E+02	−8.90E+02	−8.87E+02	−8.82E+02	−8.86E+02	**−8.94E+02**	−8.82E+02
	σ_f	2.41E+01	1.81E+01	2.25E+01	2.39E+01	1.80E+01	**1.47E+01**	2.37E+01	1.63E+01	2.07E+01	1.54E+01

Bold values correspond to the best value obtained for each function

results in 8 of the 14 functions. However, if another parameter set is used, it results in a bad performance.

Finally, in the third stage, the performance of the proposed algorithm is evaluated considering different values for γ. In the simulation, the values of γ are varied from 0.001 to 0.01 whereas the values of α_1 and α_2 remain fixed at 1.4 and 0.05, respectively. Table 3.4 summarizes the results of this experiment. From the information provided by Table 3.4, it can be seen that the proposed fuzzy algorithm with $\gamma = 0.005$ obtains the best performance on functions f_1, f_2, f_3, f_4, f_6, f_7, f_{10}, f_{12} and f_{13}. However, when the parameter γ takes any other value, the performance of the algorithm is inconsistent. Under this configuration, the algorithm presents the best possible performance, since it obtains the best indexes in 10 out of 14 functions. In general, the experimental results shown in Tables 3.2, 3.3 and 3.4 suggest that a proper combination of the parameter values can improve the performance of the proposed method and the quality of solutions. In this experiment we can conclude that the best parameter set is composed by the following values: $\alpha_1 = 1.4$, $\alpha_2 = 0.05$ and $\gamma = 0.005$.

Once the parameters α_1, α_2 and γ have been experimentally set, it is possible to analyze their influence in the optimization process. In the search strategy, integrated in the fuzzy system, α_1 modifies the attraction that a promising individual experiment with regard to the best current element in the population. This action aims to improve the solution quality of the individual, considering that the unexplored region between the promising solution and the best element could contain a better solution. On the other hand, α_2 adjusts the repulsion to which a low-quality individual is undergone. This operation intends to enhance the quality of the bad candidate solution through a movement in opposite direction of the best current element. This repulsion is considered, since there is evidence that the unexplored section between the low-quality solution and the best current element does not enclose promising solutions. Finally, γ defines the neighborhood around a promising solution, from which a local search operation is conducted. The objective of this process is to refine the quality of each solution that initially maintains an acceptable fitness value.

Considering their magnitude, the values of $\alpha_1 = 1.4$, $\alpha_2 = 0.05$ and $\gamma = 0.005$ indicate that the attraction procedure is the most important operation in the optimization strategy. This fact confirms that the attraction process represents the most prolific operation in the fuzzy strategy, since it searches new solutions in the direction where high fitness values are expected. According to its importance, the repulsion operation holds the second position. Repulsion produces significant small modifications of candidate solutions in comparison to the attraction process. This result indicates that the repulsion process involves an exploration with a higher uncertainty compared with the attraction movement. This uncertainty is a consequence of the lack of knowledge, if the opposite movement may reach a position with a better fitness value. The only available evidence is that in direction of the attraction movement, it is not possible to find promising solutions. Finally, the small value of γ induces a minor vibration for each acceptable candidate solution, in order to refine its quality in terms of fitness value.

Table 3.4 Experimental results obtained by the proposed algorithm using different values of γ

γ		0.001	0.002	0.003	0.004	0.005	0.006	0.007	0.008	0.009	0.01
f_1	$\bar{f}$	3.73E−28	5.52E−29	4.79E−29	1.14E−28	**1.01E−39**	2.04E−28	1.92E−28	1.20E−28	8.28E−29	1.27E−28
	σ_f	8.01E−28	2.74E−28	2.23E−28	5.10E−28	**5.54E−39**	7.66E−28	7.31E−28	4.91E−28	4.46E−28	6.93E−28
f_2	$\bar{f}$	1.78E−16	5.62E−16	6.03E−16	9.19E−17	**5.26E−35**	4.20E−16	2.44E−22	7.16E−22	3.97E−17	2.41E−18
	σ_f	8.72E−16	2.14E−15	2.30E−15	4.21E−16	**2.88E−34**	1.63E−15	1.34E−21	3.89E−21	2.18E−16	1.30E−17
f_3	$\bar{f}$	1.19E−23	5.91E−25	1.31E−23	1.74E−24	**2.35E−25**	4.96E−24	6.38E−22	1.91E−23	3.41E−21	7.75E−13
	σ_f	6.49E−23	2.61E−24	6.52E−23	5.07E−24	**8.29E−25**	2.66E−23	2.64E−21	1.04E−22	1.85E−20	4.24E−12
f_4	$\bar{f}$	3.34E−26	9.36E−16	5.20E−16	5.23E−16	**6.13E−19**	7.80E−18	5.46E−16	6.14E−16	7.75E−21	4.90E−16
	σ_f	1.52E−25	3.79E−15	2.85E−15	2.86E−15	**3.31E−18**	4.27E−17	2.99E−15	2.37E−15	4.24E−20	2.56E−15
f_5	$\bar{f}$	2.85E+01	2.85E+01	2.85E+01	**3.85E−04**	1.45E−01	2.85E+01	2.85E+01	2.85E+01	2.85E+01	2.85E+01
	σ_f	4.67E−02	4.29E−02	4.47E−02	**2.85E−02**	4.38E−02	4.52E−02	4.31E−02	3.96E−02	4.18E−02	4.49E−02
f_6	$\bar{f}$	1.89E−02	1.93E−02	1.60E−02	1.85E−02	**1.54E−02**	1.57E−02	2.17E−02	2.06E−02	2.15E−02	1.92E−02
	σ_f	1.27E−02	1.50E−02	7.99E−03	1.14E−02	**5.55E−03**	8.61E−03	1.45E−02	1.91E−02	1.90E−02	1.34E−02
f_7	$\bar{f}$	1.98E−03	1.76E−03	1.49E−03	1.58E−03	**1.32E−03**	1.71E−03	1.61E−03	1.95E−03	2.30E−03	1.36E−03
	σ_f	2.02E−03	1.62E−03	2.01E−03	1.56E−03	**1.03E−03**	1.38E−03	1.92E−03	2.03E−03	2.79E−03	1.10E−03
f_8	$\bar{f}$	−5.11E+03	−5.05E+03	**−5.27E+03**	−5.19E+03	−5.13E+03	−4.98E+03	−5.15E+03	−5.12E+03	−5.11E+03	−4.98E+03
	σ_f	5.20E+02	4.42E+02	5.69E+02	4.20E+02	4.77E+02	4.66E+02	4.54E+02	5.31E+02	**3.24E+02**	5.47E+02
f_9	$\bar{f}$	**1.14E−14**	6.94E+00	4.72E+00	1.07E+01	4.76E−01	6.13E+00	8.69E+00	2.16E+01	7.27E−02	1.41E+01
	σ_f	**2.75E−14**	2.55E+01	2.56E+01	4.05E+01	2.38E+00	3.18E+01	3.15E+01	5.64E+01	2.81E−01	4.31E+01
f_{10}	$\bar{f}$	1.23E−14	8.70E−15	9.06E−15	1.13E−14	**1.04E−14**	1.05E−14	9.06E−15	1.26E−14	8.47E−15	8.82E−15
	σ_f	6.29E−15	3.29E−15	2.97E−15	6.79E−15	**2.97E−15**	6.06E−15	3.82E−15	6.47E−15	5.55E−15	3.58E−15
f_{11}	$\bar{f}$	**0.00E+00**	**0.00E+00**	**0.00E+00**	**0.00E+00**	**0.00E+00**	4.78E−05	**0.00E+00**	**0.00E+00**	**0.00E+00**	**0.00E+00**

(continued)

Table 3.4 (continued)

	γ	0.001	0.002	0.003	0.004	0.005	0.006	0.007	0.008	0.009	0.01
f_{12}	$\bar{f}$	**0.00E+00**	**0.00E+00**	**0.00E+00**	**0.00E+00**	**0.00E+00**	1.85E−04	**0.00E+00**	**0.00E+00**	**0.00E+00**	**0.00E+00**
	σ_f	2.11E+00	9.59E−01	9.95E−01	7.95E−01	**9.38E−02**	8.96E−01	5.60E−01	1.08E+00	1.32E−01	6.74E−01
f_{13}	$\bar{f}$	7.38E+00	3.26E+00	3.37E+00	3.83E+00	**1.99E−02**	3.17E+00	2.48E+00	3.73E+00	1.76E−01	3.12E+00
	σ_f	3.98E−01	4.28E−01	4.12E−01	3.90E−01	**3.85E−01**	1.06E+00	3.91E−01	4.08E−01	1.14E+00	3.88E−01
f_{14}	$\bar{f}$	2.68E−01	2.24E−01	3.49E−01	1.90E−01	**1.51E−01**	3.82E+00	1.76E−01	2.24E−01	4.12E+00	2.08E−01
		−8.86E+02	−8.82E+02	−8.91E+02	−8.91E+02	−8.90E+02	−8.93E+02	−8.93E+02	**−8.97E+02**	−8.89E+02	−8.89E+02
	σ_f	2.29E+01	2.03E+01	2.22E+01	2.03E+01	**1.80E+01**	2.72E+01	1.91E+01	2.24E+01	1.97E+01	1.98E+01

Bold values correspond to the best value obtained for each function

3.5.2   Comparison with Other Optimization Approaches

In this subsection, the proposed method is evaluated in comparison with other popular optimization algorithms based on evolutionary principles. In the experiments, we have applied the fuzzy optimization algorithm to the 19 functions from Appendix A, and the results are compared to those produced by the Harmony Search (HS) method [14], the Bat (BAT) algorithm [15], the Differential Evolution (DE) [19], the Particle Swarm Optimization (PSO)method [12], the Artificial Bee Colony (ABC) algorithm [13] and the Covariance Matrix Adaptation Evolution Strategies (CMA-ES) [49]. These are considered the most popular metaheuristic algorithms currently in use [50]. In the experiments, the population size N has been configured to 50 individuals. The operation of the benchmark functions is conducted in 50 and 100 dimensions. In order to eliminate the random effect, each function is tested for 30 independent runs. In the comparison, a fixed number FN of function evaluations has been considered as a stop criterion. Therefore, each execution of a test function consists of $FN = 104 \times n$ function evaluations (where n represents the number of dimensions). This stop criterion has been decided to keep compatibility with similar works published in the literature [51–54]. For the comparison, all methods have been configured with the parameters, which according to their reported references reach their best performance. Such configurations are described as follows:

1. **HS** [14]: HCMR $= 0.7$ and *PArate* $= 0.3$.
2. **BAT** [15]: Loudness ($A = 2$), Pulse Rate ($r = 0.9$), Frequency minimum ($Q_{min} = 0$) and Frequency maximum ($Q_{max} = 1$).
3. **DE** [19]: $CR = 0.5$ and $F = 0.2$.
4. **PSO** [12]: $c1 = 2$ and $c2 = 2$; the weight factor decreases linearly from 0.9 to 0.2.
5. **ABC** [13]: limit $= 50$.
6. **CMA-ES** [47]: The source code has been obtained from the original author [55]. In the experiments, some minor changes have been applied to adapt CMA-ES to our test functions, but the main body is unaltered.
7. **FUZZY:** $\alpha_1 = 1.4$, $\alpha_2 = 0.05$ and $\gamma = 0.005$.

Several tests have been conducted for comparing the performance of the proposed fuzzy algorithm. The experiments have been divided in Unimodal functions (Table 3.17), Multimodal functions (Table 3.18) and Hybrid functions (Table 3.19).

3.5.2.1   Unimodal Test Functions

In this test, the performance of our proposed fuzzy algorithm is compared with HS, BAT, DE, PSO, CMA-ES and ABC, considering functions with only one optimum. Such functions are represented by functions f_1 to f_7 in Table A3.1. In the test, all functions have been operated in 50 dimensions ($n = 50$). The experimental results obtained from 30 independent executions are presented in Table 3.5.

Table 3.5 Minimization results of unimodal functions of Table 3.17 with $n = 50$

Unimodal functions of Table 3.17 with $n = 50$

		HS	BAT	DE	PSO	CMA-ES	ABC	FUZZY
f_1	$\bar{f}$	87035.2235	121388.021	61.1848761	4.39E+03	1.34E−11	3.09E−06	**2.30E−29**
	σ_f	5262.26532	6933.12929	163.555175	1261.19173	5.19E−12	3.44E−06	**1.17E−28**
	f_{best}	76937.413	108807.878	0.03702664	1.65E+03	5.69E−12	2.47E−07	**5.17E−114**
	f_{worst}	95804.9747	138224.113	878.436103	7.37E+03	2.55E−11	1.72E−05	**6.42E−28**
f_2	$\bar{f}$	1.37E+14	4.32E+17	0.04057031	4.54E+01	**9.92E−06**	1.39E−03	4.15E−04
	σ_f	3.19E+14	1.54E+18	0.09738928	16.386199	**2.55E−06**	0.00071159	0.00227186
	f_{best}	1.04E+10	1633259021	4.03E−12	2.61E+01	5.87E−06	5.62E−04	**7.20E−59**
	f_{worst}	1.64E+15	7.60E+18	0.45348954	9.75E+01	**1.44E−05**	2.98E−03	0.01244379
f_3	$\bar{f}$	130472.801	297342.421	55982.8182	1.57E+04	2.89E−03	4.14E+04	**1.93E−05**
	σ_f	11639.2864	99049.8321	9234.85975	9734.92204	0.00164804	4785.18216	**4.28E−05**
	f_{best}	104514.012	164628.01	36105.5799	4.23E+03	9.88E−04	2.85E+04	**1.66E−10**
	f_{worst}	147659.604	563910.174	70938.4205	4.96E+04	8.88E−03	4.84E+04	**0.00018991**
f_4	$\bar{f}$	80.1841708	90.1756477	25.8134455	2.32E+01	3.96E−04	7.35E+01	**3.37E−16**
	σ_f	2.55950002	1.86267545	6.30765469	3.51409694	8.21E−05	3.60905231	**1.85E−15**
	f_{best}	73.2799506	86.1129762	15.7894785	1.73E+01	2.57E−04	6.55E+01	**7.52E−70**
	f_{worst}	83.8375161	92.7805806	38.8210447	3.06E+01	5.65E−04	7.90E+01	**1.01E−14**
f_5	$\bar{f}$	1024.70257	276.243833	52.5359064	6.04E+02	**3.51E−05**	4.53E+01	4.85E−04
	σ_f	100.932656	45.1209564	7.69858817	198.334321	0.49723274	1.13628434	**0.0389642**
	f_{best}	783.653134	211.600116	47.1421071	289.29993	**1.21E−09**	42.1783081	3.30E−09
	f_{worst}	1211.08532	399.160851	75.1362468	1126.38574	**3.0654249**	47.7422282	4.6323356

(continued)

Table 3.5 (continued)

Unimodal functions of Table 3.17 with $n = 50$

		HS	BAT	DE	PSO	CMA-ES	ABC	FUZZY
f_6	$\bar{f}$	88027.4244	119670.641	43.5155273	4.51E+03	**1.42E−11**	4.15E−06	2.18E−07
	σ_f	5783.21576	6818.7235	80.4217558	2036.72193	**5.53E−12**	8.56E−06	0.84607249
	f_{best}	77394.5062	105958.622	0.01832758	1705.47866	**5.88E−12**	6.00E−07	1.18513127
	f_{worst}	97765.4819	130549.736	306.098587	13230.6439	**2.85E−11**	4.79E−05	5.18913374
f_7	$\bar{f}$	197.476174	116.81967	0.08164158	4.43E+01	2.82E−02	6.86E−01	**3.43E−04**
	σ_f	28.808573	16.4654239	0.12240289	17.8200508	0.00499868	0.14547266	**0.00447976**
	f_{best}	116.483527	87.6450119	0.01387586	15.7697307	0.0201694	0.41798576	**0.00018152**
	f_{worst}	263.233333	156.02459	0.65353574	85.526355	0.03888318	0.8957427	**0.02057915**

Bold values correspond to the best minima obtained for each function and for each algorithm tested

They report the averaged best fitness values $(\bar{f})$ and the standard deviations (σ_f) obtained in the runs. We have also included the best (f_{Best}) and the worst (f_{Worst}) fitness values obtained during the total number of executions. The best entries in Table 3.5 are highlighted in boldface. From Table 3.5, according to the averaged best fitness value $(\bar{f})$ index, we can conclude that the proposed method per-forms better than the other algorithms in functions f_1, f_3, f_4 and f_7. In the case of functions f_2, f_5 and f_6, the CMA-ES algorithm obtains the best results. By contrast, the rest of the algorithms presents different levels of accuracy, with ABC being the most consistent. These results indicate that the proposed approach provides better performance than HS, BAT, DE, PSO and ABC for all functions except for the CMA-ES which delivers similar results to those produced by the proposed approach. By analyzing the standard deviation (σ_f) Index in Table 3.5, it becomes clear that the metaheuristic method which presents the best results also normally obtains the smallest deviations. To statistically analyze the results of Table 3.5, a non-parametric test known as the Wilcoxon analysis [56, 57] has been conducted. It allows us to evaluate the differences between two related methods. The test is performed for the 5% (0.05) significance level over the "averaged best fitness values $(\bar{f})$" data. Table 3.6 reports the p-values generated by Wilcoxon analysis for the pair-wise comparison of the algorithms. For the analysis, five groups are produced: FUZZY versus HS, FUZZY versus BAT, FUZZY versus DE, FUZZY versus PSO, FUZZY versus CMA-ES and FUZZY versus ABC. In the Wilcoxon analysis, it is considered a null hypothesis that there is no notable difference between the two methods. On the other hand, it is admitted as an alternative hypothesis that there is an important difference between the two approaches. In order to facilitate the analysis of Table 3.6, the symbols ▲, ▼, and ▶ are adopted. ▲ indicates that the proposed method performs significantly better

Table 3.6 p-values produced by Wilcoxon test comparing FUZZY versus HS, FUZZY versus BAT, FUZZY versus DE, FUZZY versus PSO, FUZZY versus CMA-ES and FUZZY versus ABC over the "averaged best fitness values" from Table 3.5

Wilcoxon test for unimodal functions of Table 3.17 with n = 50						
FUZZY vs	HS	BAT	DE	PSO	CMA-ES	ABC
f_1	5.02E−07▲	9.50E−08▲	7.36E−07▲	7.90E−05▲	7.23E−03▲	5.15E−04▲
f_2	4.06E−07▲	2.46E−08▲	2.08E−03▲	2.02E−05▲	0.0937▶	3.04E−03▲
f_3	2.02E−08▲	3.75E−08▲	1.05E−07▲	4.16E−05▲	0.0829▶	2.76E−06▲
f_4	3.55E−07▲	2.15E−08▲	3.41E−06▲	2.01E−06▲	8.14E−03▲	4.17E−07▲
f_5	1.08E−08▲	4.05E−09▲	2.04E−07▲	8.14E−09▲	0.1264▶	1.25E−07▲
f_6	6.18E−07▲	6.55E−08▲	4.60E−06▲	2.16E−07▲	0.0741▶	2.15E−03▲
f_7	4.36E−07▲	1.92E−08▲	2.81E−04▲	5.49E−06▲	0.1031▶	1.04E−03▲
▲	7	7	7	7	2	7
▼	0	0	0	0	0	0
▶	0	0	0	0	5	0

than the tested algorithm on the specified function. ▼ symbolizes that the proposed algorithm performs worse than the tested algorithm, and ▶ means that the Wilcoxon rank sum test cannot distinguish between the simulation results of the fuzzy optimizer and the tested algorithm. The number of cases that fall in these situations are shown at the bottom of the table. After an analysis of Table 3.6, it is evident that all p-values in the FUZZY versus HS, FUZZY versus BAT, FUZZY versus DE, FUZZY versus PSO and FUZZY versus ABC columns are less than 0.05 (5% significance level) which is a strong evidence against the null hypothesis and indicates that the proposed method performs better (▲) than the HS, BAT, DE, PSO and ABC algorithms. This data is statistically significant and shows that it has not occurred by coincidence (i.e. due to the normal noise contained in the process). In the case of the comparison between FUZZY and CMA-ES, the FUZZY method maintains a better (▲) performance in functions f_1 and f_4. In functions f_2, f_3, f_5, f_6 and f_7 the CMA-ES presents a similar performance to the FUZZY method. This fact can be seen from the column FUZZY versus CMA-ES, where the p-values of functions f_2, f_3, f_5, f_6 and f_7 are higher than 0.05 (▶). These results reveal that there is no statistical difference in terms of precision between FUZZY and CMA-ES, when they are applied to the afore mentioned functions. In general, the results of the Wilcoxon analysis demonstrate that the proposed algorithm performs better than most of the other methods.

In addition to the experiments in 50 dimensions, we have also conducted a set of simulations on 100 dimensions to test the scalability of the proposed fuzzy method. In the analysis, we also employed all the compared algorithms in this test. The simulation results are presented in Tables 3.7 and 3.8, which report the data produced during the 30 executions and the Wilcoxon analysis, respectively. According to the averaged best fitness value $(\bar{f})$ index from Table 3.7, the proposed method performs better than the other algorithms in functions f_1, f_2, f_3, f_4 and f_7. In the case of functions f_5 and f_6, the CMA-ES algorithm obtains the best results. On the other hand, the rest of the algorithms present different levels of accuracy. After analyzing Table 3.7, it is clear that the proposed fuzzy method presents slightly better results than CMA-ES in 100 dimensions. From Table 3.8, it is evident that all p-values in the FUZZY versus HS, FUZZY versus BAT, FUZZY versus DE, FUZZY versus PSO and FUZZY versus ABC columns are less than 0.05, which indicates that the pro-posed method performs better than the HS, BAT, DE, PSO and ABC algorithms. In the case of FUZZY versus CMA-ES, the FUZZY method maintains a better performance in functions f_1, f_2 and f_4 . In functions f_3, f_5, f_6 and f_7 the CMA-ES presents a similar performance to the FUZZY method. This experiment shows that the more dimensions there are, the worse the performance of the CMA-ES is.

3.5.2.2 Multimodal Test Functions

Contrary to unimodal functions, multimodal functions include many local optima. For this cause, they are, in general, more complicated to optimize. In this test the performance of our algorithm is compared with HS, BAT, DE, PSO, CMA-ES and

Table 3.7 Minimization results of unimodal functions of Table 3.17 with n = 100

Unimodal functions of Table 3.17 with n = 100

		HS	BAT	DE	PSO	CMA-ES	ABC	FUZZY
f_1	$\bar{f}$	2.19E+05	2.63E+05	3.89E+02	1.43E+04	1.32E−05	2.45E−02	**1.89E−16**
	σ_f	10311.7753	14201.56	323.6077	2920.781	3.21E−06	0.025751	**1.04E−15**
	f_{best}	1.74E+05	2.30E+05	1.26E+01	9.64E+03	8.00E−06	4.46E−03	**6.43E−68**
	f_{worst}	2.30E+05	2.89E+05	1.38E+03	2.09E+04	2.04E−05	1.26E−01	**5.67E−15**
f_2	$\bar{f}$	7.24E+37	1.31E+45	5.73E−01	1.51E+02	1.26E−02	9.28E−02	**1.89E−08**
	σ_f	2.59E+38	6.91E+45	0.577756	41.12351	0.002871	0.025454	**1.03E−07**
	f_{best}	5.45E+31	3.36E+34	2.94E−02	9.05E+01	8.91E−03	5.40E−02	**4.84E−46**
	f_{worst}	1.35E+39	3.79E+46	2.56E+00	2.52E+02	2.38E−02	0.183538	**5.67E−07**
f_3	$\bar{f}$	4.98E+05	1.15E+06	2.84E+05	7.90E+04	8.76E−04	1.84E+05	**2.07E−08**
	σ_f	58467.2769	312595.4	27133	34174.72	0.774352	20108.58	**0.548998**
	f_{best}	3.37E+05	4.69E+05	2.31E+05	3.71E+04	8.76E−04	1.27E+05	**3.94E−09**
	f_{worst}	616974.994	1942096	356515.6	160140	145362.6	219982.4	**2.251599**
f_4	$\bar{f}$	9.01E+01	9.46E+01	4.05E+01	2.81E+01	2.16E−06	9.04E+01	**2.63E−15**
	σ_f	1.11508888	0.859592	6.841909	3.258179	0.039821	1.829735	**6.06E−15**
	f_{best}	8.69E+01	9.29E+01	2.70E+01	2.15E+01	1.39E−08	8.37E+01	**1.24E−67**
	f_{worst}	9.16E+01	9.62E+01	5.55E+01	3.40E+01	2.91E−01	9.30E+01	**2.02E−14**
f_5	$\bar{f}$	2.70E+03	1.15E+03	1.29E+02	3.95E+03	**9.09E−04**	1.04E+02	9.86E−04
	σ_f	540.831375	88.37934	18.77995	641.937	1.455757	6.275118	**0.145843**
	f_{best}	0	980.5006	101.8736	2571.715	**1.32E−05**	96.95089	2.43E−06
	f_{worst}	3247.18778	1308.545	167.9735	5316.764	**94.82761**	121.3717	99.20371
f_6	$\bar{f}$	2.21E+05	2.64E+05	3.70E+02	1.45E+04	**1.51E−05**	1.92E−02	2.02E−05
	σ_f	9381.48118	18216.59	280.554	2798.891	**2.27E−06**	0.018164	2.874131
	f_{best}	198863.662	226296	16.80234	9243.441	**1.10E−05**	0.002508	1.50E−06
	f_{worst}	235307.769	288557.5	1278.669	20954.27	**1.94E−05**	0.091202	23.74369
f_7	$\bar{f}$	1.28E+03	4.67E+02	7.35E−01	4.85E+02	7.03E−02	2.20E+00	**4.51E−03**
	σ_f	100.811769	54.79476	0.588307	125.2015	0.01044	0.36193	**0.005359**
	f_{best}	975.480173	351.3869	0.084009	233.7028	0.046899	1.43279	**3.59E−05**
	f_{worst}	1424.35137	609.1208	2.383158	806.3506	0.08989	2.987775	**0.021358**

Bold values correspond to the best minima obtained for each function and for each algorithm tested

ABC regarding multimodal functions. Multimodal functions are represented by functions from f_8 to f_{14} in Table 3.18, where the number of local minima increases exponentially as the dimension of the function increases. Under such conditions, the experiment reflects the ability of each algorithm to find the global optimum in the presence of numerous local optima. In the simulations, the functions are operated in 50 dimensions ($n = 50$). The results, averaged over 30 executions, are reported in Table 3.9 in terms of the best fitness values $\left(\bar{f}\right)$ and the standard deviations $\left(\sigma_f\right)$.

Table 3.8 p-values produced by Wilcoxon test comparing FUZZY versus HS, FUZZY versus BAT, FUZZY versus DE, FUZZY versus PSO, FUZZY versus CMA-ES and FUZZY versus ABC over the "averaged best fitness values" from Table 3.7

Wilcoxon test for unimodal functions of Table 3.17 with n = 150						
FUZZY versus	HS	BAT	DE	PSO	CMA-ES	ABC
f_1	3.02E−08▲	8.08E−08▲	6.49E−06▲	2.01E−07▲	7.95E−04▲	4.02E−05▲
f_2	5.35E−11▲	1.13E−11▲	7.50E−05▲	4.70E−06▲	2.49E−04▲	8.29E−04▲
f_3	6.01E−07▲	7.07E−08▲	4.68E−07▲	8.47E−06▲	7.43E−02▶	2.30E−07▲
f_4	1.49E−07▲	3.79E−07▲	2.01E−06▲	1.50E−06▲	7.47E−04▲	2.20E−07▲
f_5	8.79E−06▲	5.50E−06▲	9.47E−05▲	2.16E−07▲	1.85E−01▶	4.72E−05▲
f_6	2.73E−07▲	4.79E−07▲	8.05E−05▲	5.79E−06▲	2.45E−01▶	1.49E−04▲
f_7	1.05E−07▲	5.42E−06▲	2.01E−04▲	7.62E−06▲	8.51E−02▶	4.61E−05▲
▲	7	7	7	7	3	7
▼	0	0	0	0	0	0
▶	0	0	0	0	4	0

The best results are highlighted in boldface. Likewise, p-values of the Wilcoxon test of 30 independent repetitions are exhibited in Table 3.10. In the case of f_8, f_{10}, f_{11} and f_{14}, the proposed fuzzy method presents a better performance than HS, BAT, DE, PSO, CMA-ES and ABC. For functions f_{12} and f_{13}, the fuzzy approach exhibits a worse performance compared to CMA-ES. Additionally, in the case of function f_9 the proposed method and ABC maintain the best performance compared to HS, BAT, DE, PSO and CMA-ES. The rest of the algorithms present different levels of accuracy, with ABC being the most consistent. In particular, this test yields a large difference in performance, which is directly related to a better trade-off between exploration and exploitation produced by the formulated rules of the proposed fuzzy method.

The results of the Wilcoxon analysis, presented in Table 3.10, statistically demonstrate that the proposed algorithm performs better than HS, DE, BAT, DE and PSO in all test functions ($f_8 - f_{14}$). In the case of the comparison between FUZZY and CMA-ES, the FUZZY method maintains a better (▲) performance in functions f_8, f_9, f_{10}, f_{11} and f_{14}. On the other hand, in functions f_{12} and f_{13} the FUZZY method presents worse results (▼) than the CMA-ES algorithm. However, according to Table 3.10, the proposed FUZZY approach obtains a better performance than ABC in all cases except for function f_9, where there is no difference in results between the two.

In addition to the 50 dimensions benchmark function tests, we also performed a series of simulations with 100 dimensions by using the same set of functions in Table 3.18. The results are presented in Tables 3.11 and 3.12, which report the data produced during the 30 executions and the Wilcoxon analysis, respectively. In Table 3.11, it can be seen that the proposed method performs better than HS, BAT, DE, PSO, CMA-ES and ABC for functions f_8, f_9, f_{10}, f_{11} and f_{13}. On the other

Table 3.9 Minimization results of multimodal functions of Table 3.18 with n = 50

		HS	BAT	DE	PSO	CMA-ES	ABC	FUZZY
f_8	$\bar{f}$	−5415.83905	−3270.254967	−20232.0393	−1.00E+04	−6.22E+03	−1.94E+04	**−2.69E+05**
	σ_f	**318.326084**	474.2974644	799.670519	1139.72504	577.603827	328.074723	338131.298
	f_{best}	−6183.25306	−4253.007751	−20830.6009	−12207.27	−7910.14987	−20460.0202	**−1602802.18**
	f_{worst}	−4937.01673	−2560.016959	−16849.3789	−7417.20923	−5277.36121	−18598.5718	**−51860.8124**
f_9	$\bar{f}$	637.314967	370.181231	94.9321639	2.78E+02	0.00912	**0.00000643**	0.00000103
	σ_f	25.9077403	31.0789956	23.8913991	38.1965572	8.314267	**2.34825349**	1.6781696
	f_{best}	581.055495	321.398632	49.5787092	2.09E+02	0.000691	1.99899276	**0**
	f_{worst}	681.505155	450.919651	143.664199	375.979507	342.621828	11.2277505	310.43912
f_{10}	$\bar{f}$	20.2950743	19.2995398	1.04072229	1.21E+01	8.74E−07	0.0169	**1.4E−14**
	σ_f	0.09866263	0.11146929	0.69779278	1.03376556	1.4921E−07	0.01136192	**4.489E−15**
	f_{best}	19.9003135	18.9741996	0.0040944	9.35E+00	5.62E−07	0.00292	**7.99E−15**
	f_{worst}	20.472367	19.576839	2.78934934	1.38E+01	0.0000118	0.053	**2.22E−14**
f_{11}	$\bar{f}$	786.564993	1072.40695	0.98725915	4.05E+01	9.87E−10	0.00623	**0.00E+00**
	σ_f	49.0195978	70.0220465	0.62998733	12.8397453	4.8278E−10	0.01154936	0
	f_{best}	658.158623	926.062051	0.00107768	14.1594978	2.92E−10	0.01503879	0
	f_{worst}	860.983823	1186.93137	2.4153938	64.7187195	2.32E−09	0.053391	0
f_{12}	$\bar{f}$	557399404	1029876322	1309.87126	4.08E+01	**2.58E−12**	3.67E−07	1.91E−08
	σ_f	68320767.6	150067294	4319.40539	27.0146375	**1.0706E−12**	4.1807E−07	0.63138873
	f_{best}	444164964	763229039	0.20366113	16.607083	**9.05E−13**	2.14E−08	1.95E−08
	f_{worst}	700961313	1277767934	17508.9826	136.891908	**5.63E−12**	0.00000184	19.8206283
f_{13}	$\bar{f}$	1163989772	1982187734	29551.4297	1.41E+05	**5.03E−11**	0.00000598	9.06E−09

(continued)

Table 3.9 (continued)

		HS	BAT	DE	PSO	CMA-ES	ABC	FUZZY
	σ_f	123421334	291495991	137415.981	186740.994	**2.7928E−11**	7.5042E−06	0.0439066
	f_{best}	898858903	1250159582	4.37613356	1594.63864	**1.37E−11**	5.06E−07	6.91E−10
	f_{worst}	1453640537	2558128052	754699.3	735426.524	**1.38E−10**	0.0000348	60.2604938
f_{14}	$\bar{f}$	−958.679663	−1066.80779	−1937.10075	−1.40E+03	−1.84E+03	−1.32E+03	**−1.96E+03**
	σ_f	40.8885038	61.5793102	13.199329	77.8992722	41.6063335	29.4904745	**0.06633944**
	f_{best}	−1060.29598	−1213.46466	−1958.29881	−1527.69665	−1915.89813	−1395.05118	**−1958.30818**
	f_{worst}	−893.619964	−988.87492	−1915.09727	−1275.90292	−1732.12078	−1262.8218	**−1958.04305**

Bold values correspond to the best value obtained for each function

Table 3.10 p-values produced by Wilcoxon test comparing FUZZY versus HS, FUZZY versus BAT, FUZZY versus DE, FUZZY versus PSO, FUZZY versus CMA-ES and FUZZY versus ABC over the "averaged best fitness values" values from Table 3.9

Wilcoxon test for unimodal functions of Table 3.17 with $n = 150$

FUZZY vs	HS	BAT	DE	PSO	CMA-ES	ABC
f_8	7.14E−05▲	5.40E−05▲	3.48E−05▲	1.15E−05▲	4.72E−06▲	2.50E−05▲
f_9	4.16E−05▲	2.49E−05▲	6.15E−04▲	2.12E−05▲	4.29E−04▲	0.0783▶
f_{10}	3.14E−05▲	3.02E−05▲	7.49E−04▲	1.50E−05▲	3.12E−04▲	9.49E−04▲
f_{11}	5.50E−08▲	9.33E−09▲	7.14E−04▲	5.38E−06▲	8.50E−03▲	6.15E−03▲
f_{12}	6.48E−11▲	8.40E−11▲	4.68E−08▲	5.29E−06▲	7.24E−04▼	4.03E−03▲
f_{13}	7.98E−11▲	9.79E−11▲	4.16E−10▲	7.47E−11▲	9.40E−04▼	4.55E−05▲
f_{14}	7.14E−06▲	4.58E−07▲	5.79E−04▲	8.16E−05▲	9.64E−04▲	5.68E−05▲
▲	7	7	7	7	5	6
▼	0	0	0	0	2	0
▶	0	0	0	0	0	1

hand, the CMA-ES maintains better results than HS, BAT, DE, PSO, ABC and the fuzzy optimizer for function f_{12}. Likewise, the DE method obtains better indexes than the other algorithms for function f_{14}. From the Wilcoxon analysis shown in Table 3.12, the results indicate that the proposed method performs better than the HS, BAT, DE, PSO and ABC algorithms. In the case of FUZZY versus CMA-ES, the FUZZY method maintains a better performance in all test functions except in problem f_{12}, where the CMA-ES produces better results than the proposed FUZZY method. This experiment also shows that the more dimensions there are, the worse the performance of the CMA-ES is.

3.5.2.3 Hybrid Test Functions

In this test, hybrid functions are employed to test the optimization performance of the proposed approach. Hybrid functions, shown in Table A3.3, are multimodal functions with complex behaviors, since they are built from different multimodal single functions. A detailed implementation of the hybrid functions can be found in [46]. In the experiments, the performance of our proposed fuzzy algorithm is compared with HS, BAT, DE, PSO, CMA-ES and ABC, considering functions f_{15} to f_{19}. In the first test, all functions have been operated in 50 dimensions ($n = 50$). The experimental results obtained from 30 independent executions are presented in Tables 3.13 and 3.14. In Table 3.13, the indexes $\bar{f}$, σ_f, f_{Best} and f_{Worst}, obtained during the total number of executions, are reported. Furthermore, Table 3.14 presents the statistical Wilcoxon analysis of the averaged best fitness values $\bar{f}$ from Table 3.13. According to Table 3.13, the proposed approach maintains a superior performance in comparison to most of the other methods. In the case of f_{15}, f_{16} and f_{18}, the proposed

Table 3.11 Minimization results of multimodal functions from Table 3.18 with $n = 100$

Multimodal functions of Table 3.18 with $n = 100$

		HS	BAT	DE	PSO	CMA-ES	ABC	FUZZY
f_8	$\bar{f}$	−7.87E+03	−4.47E+03	−2.34E+04	−1.48E+04	−8.66E+03	−3.51E+04	**−1.62E+05**
	σ_f	584.340059	799.31231	3914.77561	1729.15916	831.547363	**554.101741**	47514.2941
	f_{best}	−9141.81789	−6497.01004	−32939.624	−19362.7097	−10177.5382	−36655.8271	**−272938.668**
	f_{worst}	−6669.02042	−2858.38997	−18358.9463	−12408.7377	−7375.40434	−33939.9476	**−87985.343**
f_9	$\bar{f}$	1.44E+03	9.12E+02	3.98E+02	7.49E+02	2.37E+02	6.49E+01	**4.00E−05**
	σ_f	41.464352	60.4785656	67.0536747	67.5932069	209.403173	7.90879809	**0.00015207**
	f_{best}	1293.19918	772.978721	205.209989	635.167862	74.8466353	49.9922666	**0**
	f_{worst}	1503.0147	1033.80416	522.504376	865.893944	806.576683	79.9136576	**0.0005994**
f_{10}	$\bar{f}$	2.06E+01	1.98E+01	2.42E+00	1.33E+01	6.14E−04	3.01E+00	**3.96E−12**
	σ_f	0.06052521	0.08670819	0.82882063	0.87767923	9.62E−05	0.31154878	**2.15E−11**
	f_{best}	2.05E+01	1.96E+01	1.12E+00	1.16E+01	4.14E−04	2.26E+00	**1.51E−14**
	f_{worst}	2.08E+01	2.00E+01	4.63E+00	1.51E+01	8.49E−04	3.59E+00	**1.18E−10**
f_{11}	$\bar{f}$	1.96E+03	2.38E+03	4.46E+00	1.18E+02	1.20E−03	1.61E−01	**0.00E+00**
	σ_f	81.7177655	134.986806	2.67096611	22.637086	0.00028091	0.14553457	**0**
	f_{best}	1773.68789	2035.67623	1.02086153	86.1907763	0.00066065	0.01309143	**0**
	f_{worst}	2083.84569	2557.11279	12.932624	170.763675	2.25E−03	0.64617359	**0**
f_{12}	$\bar{f}$	1.88E+09	2.55E+09	4.85E+04	2.06E+04	**2.20E−06**	5.96E−03	2.67E+02
	σ_f	110307213	242921094	140231.648	50859.5682	**7.19E−07**	0.01533206	1423.14186
	f_{best}	1.64E+09	1.87E+09	1.86E+00	2.93E+01	**1.27E−06**	2.91E−05	4.21E−01
	f_{worst}	2090764763	2966731722	769582.376	251450.989	**4.07E−06**	0.06229125	7801.38816

(continued)

Table 3.11 (continued)

Multimodal functions of Table 3.18 with $n = 100$

		HS	BAT	DE	PSO	CMA-ES	ABC	FUZZY
f_{13}	$\bar{f}$	3.54E+09	4.83E+09	6.63E+05	1.77E+06	2.20E+01	4.74E−03	**4.71E−05**
	σ_f	255789666	477261470	1076152.4	1373210.26	33.1622741	0.00582626	**1.25E−05**
	f_{best}	2936347707	3998731504	3147.16418	417561.443	9.98142144	0.00075547	**2.65E−05**
	f_{worst}	3928279153	5901118241	4651437.83	7741055.37	176.314279	0.02480097	**7.86E−05**
f_{14}	$\bar{f}$	−1.57E+03	−1.82E+03	**−3.83E+03**	−2.46E+03	−3.57E+03	−3.78E+03	−2.30E+03
	σ_f	79.1693639	76.7852589	36.6930452	97.0301039	59.5947211	**22.8858428**	65.2812493
	f_{best}	−1722.36681	−1945.12523	**−3900.49741**	−2683.50044	−3676.29234	−3823.95029	−2455.01558
	f_{worst}	−1432.17812	−1695.67279	**−3740.63055**	−2280.67655	−3393.55796	−3724.2658	−2156.74064

Bold values correspond to the best minima obtained for each function and for each algorithm tested

Table 3.12 p-values produced by Wilcoxon test comparing FUZZY versus HS, FUZZY versus BAT, FUZZY versus DE, FUZZY versus PSO, FUZZY versus CMA-ES and FUZZY versus ABC over the "averaged best fitness values" from Table 3.11

Wilcoxon test for multimodal functions of Table 3.17 with $n = 100$						
FUZZY vs	HS	BAT	DE	PSO	CMA-ES	ABC
f_8	8.13E−05▲	6.47E−05▲	4.69E−05▲	3.17E−05▲	7.12E−05▲	3.79E−05▲
f_9	1.36E−07▲	9.50E−06▲	7.60E−06▲	8.66E−06▲	5.49E−06▲	3.14E−06▲
f_{10}	6.95E−07▲	6.35E−07▲	3.57E−06▲	6.13E−06▲	3.17E−04▲	3.93E−06▲
f_{11}	4.98E−07▲	8.16E−07▲	7.14E−04▲	5.31E−06▲	2.02E−03▲	9.49E−03▲
f_{12}	4.27E−08▲	7.68E−08▲	8.47E−07▲	7.47E−07▲	3.05E−06▼	1.64E−05▲
f_{13}	4.59E−09▲	6.47E−09▲	6.17E−07▲	7.47E−08▲	9.17E−06▲	7.47E−04▲
f_{14}	8.16E−05▲	8.96E−05▲	6.49E−04▼	9.42E−03▲	5.47E−04▲	6.01E−04▲
▲	7	7	6	7	6	7
▼	0	0	1	0	1	0
►	0	0	0	0	0	0

fuzzy method performs better than HS, BAT, DE, PSO, CMA-ES and ABC. For function f_{19}, the fuzzy approach presents a worst performance than CMA-ES or ABC. However, in functions f_{16} and f_{18}, the proposed method and ABC maintain a better performance than HS, BAT, DE, PSO and CMA-ES. For function f_{17} the proposed FUZZY method and CMA-ES perform better than other methods. Therefore, the proposed fuzzy algorithm reaches better $\bar{f}$ values in 4 from 5 different functions. This fact confirms that the fuzzy method is able to produce more accurate solutions than its competitors. From the analysis of σ_f in Table 3.13, it is clear that our fuzzy method obtains a better consistency than the other algorithms, since its produced solutions present a small dispersion. As it can be expected, the only exception is function f_{19}, where the fuzzy algorithm does not achieve the best performance Additional to suchresults, Table 3.13 shows that the proposed Fuzzy method attains the best produced solution f_{Best} during the 30 independent executions than the other algorithms, except for f_{19} function. Besides, the worst fitness values f_{Worst} generated by the fuzzy technique maintain a better solution quality than the other methods excluding function f_{19}. The case of obtaining the best f_{Best} and f_{Worst} indexes reflexes the remarkable capacity of the proposed fuzzy method to produce better solutions through use an efficient search strategy. Table 3.14 shows the results of the Wilcoxon analysis over the averaged best fitness values $\bar{f}$ from Table 3.13. They indicate that theproposed method performs better than the HS, BAT, DE and PSO algorithms. In the case of FUZZY versus CMA-ES, the FUZZY method maintains a better performance in all test functions except in problem f_{19}, where the CMA-ES produces better results than the proposed FUZZY method. However, in the comparison between the FUZZY algorithm and ABC, FUZZY obtains the best results in all test functions except in problems f_{18} and f_{16}, where there is no statistical difference between the two methods.

Table 3.13 Minimization results of hybrid functions from Table 3.19 with $n = 50$

Hybrid functions of Table 3.19 with $n = 50$

		HS	BAT	DE	PSO	CMA-ES	ABC	FUZZY
f_{15}	$\bar{f}$	7.997E+13	5.102E+21	1.235E+01	9.640E+03	6.360E−06	5.230E−04	**2.490E−15**
	σ_f	1.387E+14	2.795E+22	1.692E+01	5.195E+03	2.092E−06	1.827E−04	**6.651E−15**
	f_{best}	1.531E+10	8.192E+12	1.190E−02	4.190E+03	4.160E−06	2.680E−04	**3.170E−58**
	f_{worst}	4.763E+14	1.530E+23	6.581E+01	2.500E+04	1.480E−05	1.020E−03	**2.530E−14**
f_{16}	$\bar{f}$	2.707E+03	3.508E+03	7.370E+01	5.990E+02	5.880E+02	4.910E+01	**4.900E+01**
	σ_f	1.033E+02	2.525E+02	1.392E+01	1.146E+02	9.039E+00	1.997E−01	**2.581E−04**
	f_{best}	2.492E+03	2.883E+03	4.900E+01	3.938E+02	**4.900E+01**	4.900E+01	4.900E+01
	f_{worst}	2.876E+03	3.869E+03	1.037E+02	8.790E+02	8.107E+01	4.969E+01	**4.900E+01**
f_{17}	$\bar{f}$	1.151E+09	2.106E+09	6.741E+04	3.130E+05	**5.400E+01**	8.960E+02	**5.400E+01**
	σ_f	1.136E+08	1.902E+08	1.933E+05	4.214E+05	2.002E−04	1.322E+02	**9.886E−05**
	f_{best}	9.097E+08	1.637E+09	4.131E+02	1.553E+04	5.400E+01	5.467E+02	**5.400E+01**
	f_{worst}	1.429E+09	2.453E+09	9.364E+05	1.948E+06	5.400E+01	1.156E+03	**5.400E+01**
f_{18}	$\bar{f}$	2.016E+14	3.343E+19	5.426E+01	9.060E+02	5.990E+01	**4.900E+01**	**4.900E+01**
	σ_f	3.907E+14	1.830E+20	1.050E+01	2.911E+02	1.241E+01	1.211E−02	**0.000E+00**
	f_{best}	9.111E+08	3.274E+13	4.900E+01	5.306E+02	4.900E+01	4.900E+01	**4.900E+01**
	f_{worst}	1.850E+15	1.000E+21	9.712E+01	1.597E+03	8.601E+01	4.906E+01	**4.900E+01**
f_{19}	$\bar{f}$	7.742E+14	7.717E+18	−19.5833354	1.170E+06	**−1.440E+02**	−1.430E+02	2.180E+01
	σ_f	1.676E+15	4.220E+19	1.179E+02	5.836E+06	3.973E−01	3.000E−01	4.726E+02
	f_{best}	1.343E−09	1.939E+10	−143.748394	4.512E+03	**−144.056723**	−143.608756	−83.2609165
	f_{worst}	7.530E+15	2.310E+20	3.348E+02	3.205E+07	**−142.208256**	−143.0037	2.524E+03

Bold values correspond to the best minima obtained for each function and for each algorithm tested

Table 3.14 p-values produced by Wilcoxon test comparing FUZZY versus HS, FUZZY versus BAT, FUZZY versus DE, FUZZY versus PSO, FUZZY versus CMA-ES and FUZZY versus ABC over the "averaged best fitness values" from Table 3.13

Wilcoxon test for Hybrid functions of Table 3.19 with $n = 50$

FUZZY vs	HS	BAT	DE	PSO	CMA-ES	ABC
f_{15}	4.61E−10▲	7.68E−11▲	8.13E−07▲	3.17E−09▲	1.35E−04▲	5.69E−05▲
f_{16}	5.69E−08▲	6.50E−09▲	6.31E−04▲	8.43E−05▲	6.49E−05▲	1.56E−01▶
f_{17}	8.65E−11▲	9.47E−11▲	6.34E−09▲	7.35E−10▲	9.56E−02▶	4.65E−04▲
f_{18}	3.50E−11▲	7.69E−12▲	4.68E−04▲	5.31E−06▲	4.82E−04▲	1.99E−01▶
f_{19}	7.63E−10▲	9.31E−11▲	4.33E−07▲	6.00E−09▲	6.33E−07▲	5.89E−07▼
▲	5	5	5	5	4	2
▼	0	0	0	0	0	2
▶	0	0	0	0	1	1

In addition to the test in 50 dimensions, a second set of experiments have also conducted in 100 dimensions considering the same set of hybrid functions. Tables 3.15 and 3.16 present the results of the analysis in 100 dimensions. In Table 3.15, the indexes $\bar{f}$, σ_f, f_{Best} and f_{Worst}, obtained during the total number of executions, are reported. On the other hand, Table 3.14 presents the statistical Wilcoxon analysis of the averaged best fitness values $\bar{f}$ from Table 3.15.

Table 3.15 confirms the advantage of the proposed method over HS, BAT, DE, PSO, CMA-ES and ABC. After analyzing the results, it is clear that the proposed fuzzy method produces better results than HS, BAT, DE, PSO, CMA-ES and ABC in functions $f_{15} - f_{18}$. However, it can be seen that the proposed method performs worse than CMA-ES and ABC in function f_{19}. Similar to the case of 50 dimensions, in the experiments of 100 dimensions, the proposed fuzzy algorithm obtains solutions with the smallest level of dispersion (σ_f). This consistency is valid for all functions, except for problem f_{19}, where the CMA-ES obtain the best σ_f value. Considering the f_{Best} and f_{Worst} indexes, similar conclusion can be established that in the case of 50 dimensions. In 100 dimensions, it is also observed that the proposed fuzzy technique surpass all algorithms in the production of high quality solutions.

On the other hand, the data obtained from the Wilcoxon analysis (Table 3.16) demonstrates that the proposed FUZZY method performs better than the other metaheuristic algorithms in all test functions, except in problem f_{18}, where the CMA-ES and ABC produce the best results. In Table 3.16, it is also summarized the results of the analysis through the symbols ▲, ▼, and ▶. The conclusions of the Wilcoxon test statistically validate the results of Table 3.15. They indicate that the superior performance of the fuzzy method is as a consequence of a better search strategy and not for random effects.

Table 3.15 Minimization results of hybrid functions from Table 3.19 with $n = 100$

Hybrid functions of Table 3.19 with $n = 50$

		HS	BAT	DE	PSO	CMA-ES	ABC	FUZZY
f_{15}	$\bar{f}$	1.07E+38	1.45E+46	1.67E+02	3.58E+04	8.48E−03	8.13E−02	**1.03E−05**
	σ_f	2.66E+38	7.93E+46	212.77092	19777.8798	0.00248591	0.04144673	**5.64E−05**
	f_{best}	1.88E+28	1.31E+37	5.17E+00	1.58E+04	5.99E−03	4.35E−02	**4.50E−44**
	f_{worst}	1.21E+39	4.34E+47	1.15E+03	9.15E+04	1.75E−02	2.66E−01	**3.09E−04**
f_{16}	$\bar{f}$	6.46E+03	8.00E+03	1.90E+02	1.36E+03	1.55E+02	1.81E+02	**9.90E+01**
	σ_f	316.896114	426.761823	28.0273818	129.05773	18.161094	18.5185506	**0.00187142**
	f_{best}	5753.69747	7125.84623	148.521771	1116.8234	122.010095	129.807203	**98.9958572**
	f_{worst}	6997.63377	8965.08163	260.542179	1681.54333	187.628218	206.634292	**99.0057217**
f_{17}	$\bar{f}$	3.57E+09	5.03E+09	6.85E+05	2.10E+06	5.95E+02	3.72E+03	**1.09E+02**
	σ_f	251070990	401513619	1158365.05	1426951.79	84.33472	459.761456	**0.0026802**
	f_{best}	2908492728	4028081811	4767.47553	428429.323	428.761773	2973.30401	**108.99997**
	f_{worst}	3953742523	5605713616	4789383.94	6274361.22	784.972289	4756.54052	**109.01469**
f_{18}	$\bar{f}$	3.30E+38	5.29E+43	1.33E+02	2.09E+03	1.49E+02	3.34E+02	**1.08E+02**
	σ_f	1.20E+39	2.67E+44	20.7615151	481.839456	23.6851947	908.915277	**8.76059629**
	f_{best}	5.02E+29	4.38E+34	1.01E+02	1.24E+03	1.12E+02	**9.90E−01**	99.6507148
	f_{worst}	5.86E+39	1.47E+45	1.92E+02	3.51E+03	2.10E+02	4.24E+03	**134.545048**
f_{19}	$\bar{f}$	1.01E+38	1.49E+44	9.38E+03	6.45E+07	**−2.94E+02**	−2.01E+02	4.29E+07
	σ_f	3.99E+38	6.24E+44	40545.7303	217095412	**0.55047399**	1.27097202	200132558
	f_{best}	4.34E+29	1.43E+36	−1.10E+02	4.71E+04	−2.95E−02	**−295.68221**	−9.45E+01
	f_{worst}	2.18E+39	3.36E+45	2.23E+05	1.17E+09	**−2.92E+02**	−288.145624	1.08E+09

Bold values correspond to the best minima obtained for each function and for each algorithm tested

Table 3.16 p-values produced by Wilcoxon test comparing FUZZY versus HS, FUZZY versus BAT, FUZZY versus DE, FUZZY versus PSO, FUZZY versus CMA-ES and FUZZY versus ABC over the "averaged best fitness values" from Table 3.15

Wilcoxon test for Hybrid functions of Table 3.19 with $n = 100$.

FUZZY vs	HS	BAT	DE	PSO	CMA-ES	ABC
f_{15}	8.47E−12▲	9.76E−12▲	6.50E−07▲	7.00E−08▲	3.13E−04▲	7.68E−04▲
f_{16}	7.63E−05▲	8.42E−05▲	6.50E−04▲	2.00E−04▲	3.96E−04▲	6.00E−11▲
f_{17}	5.34E−08▲	6.89E−08▲	3.30E−07▲	9.04E−07▲	4.30E−04▲	8.63E−05▲
f_{18}	4.93E−12▲	8.37E−12▲	5.63E−04▲	3.46E−05▲	6.03E−04▲	1.30E−05▲
f_{19}	6.92E−11▲	2.50E−12▲	6.33E−06▲	2.02E−04▲	6.30E−07▼	4.13E−07▼
▲	5	5	5	5	4	4
▼	0	0	0	0	1	1
▶	0	0	0	0	0	0

3.5.2.4   Convergence Experiments

The comparison of the final fitness value cannot completely describe the searching performance of an optimization algorithm. Therefore, in this section, a convergence test on the seven compared algorithms has been conducted. The purpose of this experiment is to evaluate the velocity with which a compared method reaches the optimum. In the experiment, the performance of each algorithm is considered over all functions ($f_1 - f_{19}$) from Appendix A, operated in 50 dimensions. In order to build the convergence graphs, we employ the raw simulation data generated in Sects. 3.5.2.1, 3.5.2.2 and 3.5.2.3. As each function is executed 30 times for each algorithm, we select the convergence data of the run which represents the median final result. Figures 3.5, 3.6 and 3.7 show the convergence data of the seven compared algorithms. Figure 3.5 presents the convergence results for functions $f_1 - f_6$, Fig. 3.6 for functions $f_7 - f_{12}$ and Fig. 3.7 for functions $f_{13} - f_{19}$. In the figures, the x-axis is the elapsed function evaluations, and the y-axis represents the best fitness values found.

From Fig. 3.5, it is clear that the proposed fuzzy method presents a better convergence than the other algorithms for functions f_1, f_2, f_4 and f_5. However, for function f_3 and f_6 the CMA-ES reaches faster an optimal value. After an analysis of Fig. 3.5, we can say that the proposed Fuzzy method and the CMA-ES algorithm attain the best convergence responses whereas the other techniques maintain slower responses. In Fig. 3.6, the convergence graphs show that the proposed fuzzy method obtains the best responses for functions f_9, f_{10} and f_{11}. In function f_7, even though the Fuzzy technique finds in a fat way optimal solution, the DE algorithm presents the best convergence result. An interesting case is function f_9, where several optimization methods such as FUZZY, CMA-ES, ABC and DE obtain an acceptable convergence response. In case of function f_8, the DE and ABC methods own the best convergence properties. Finally, in function f_{12}, the CMA-ES attains the fastest reaction. Finally,

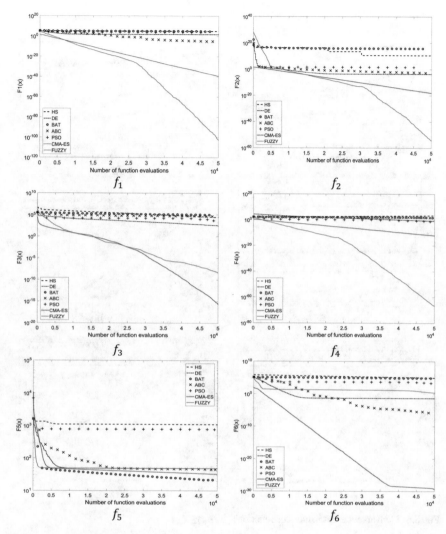

Fig. 3.5 Convergence test results for functions $f_1 - f_6$

in Fig. 3.7, the convergence responses for functions $f_{13} - f_{18}$ are presented. In function f_{13} of Fig. 3.7, the algorithms CMA-ES and ABC obtain the best responses. In case of function f_{14}, DE and ABC find an optimal solution in a prompt way than the other optimization techniques. Although for functions $f_{15} - f_{18}$ the proposed fuzzy algorithm reaches the fastest convergence reaction, the CMA-ES method maintains a similar response. For function f_{18}, the CMA-ES and ABC own the best convergence properties. Therefore, the convergence speed of the proposed fuzzy method in solving unimodal optimization ($f_1 - f_7$) problems is faster than HS, BAT, DE, PSO, CMA-ES and ABC, except in f_7, where the CMA-ES reaches the best response.

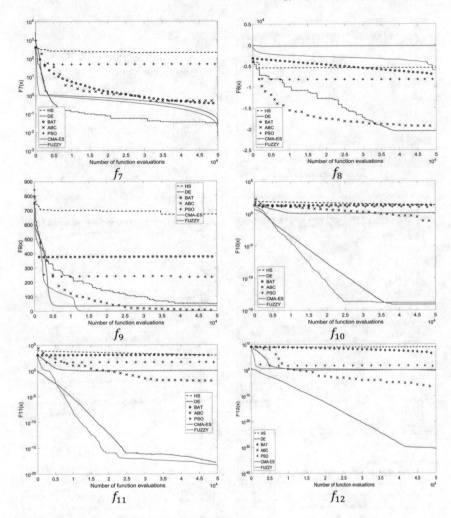

Fig. 3.6 Convergence test results for functions $f_7 - f_{12}$

On the other hand, when solving multimodal optimization problems ($f_8 - f_{14}$), the fuzzy algorithm generally converges as fast as or even faster than the com-pared algorithms. This phenomenon can be clearly observed in Figs. 3.6 and 3.7, where the proposed method generates a similar convergence curve to the others, even in the worst-case scenario. Finally, after analyzing the performance of all algorithms on hybrid functions ($f_{15} - f_{18}$), it is clear that the convergence response of the proposed approach is not as fast as CMA-ES. In fact, the proposed fuzzy and the CMA-ES algorithms present the best convergence properties when they face the optimization of hybrid functions.

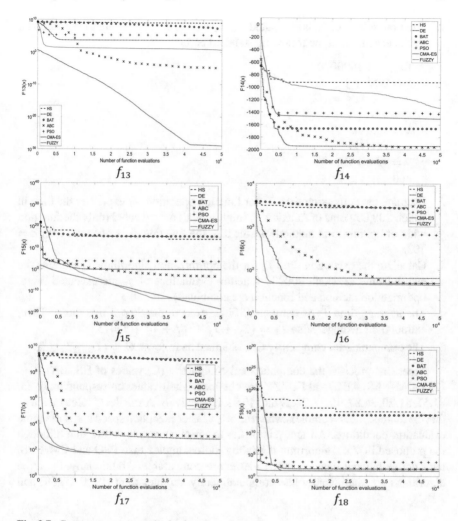

Fig. 3.7 Convergence test results for functions $f_{13} - f_{18}$

3.5.2.5 Computational Complexity

The computational complexity of all methods is evaluated. Metaheuristic methods are, in general, complex processes with several random operations and stochastic sub-routines. Therefore, it is impractical to conduct a complexity analysis from a deterministic point of view. For that reason, the computational complexity (C) is used in order to evaluate the computational effort of each algorithm. C exhibits the averaged CPU time invested by an algorithm with regard to a common time reference, when it is under operation. In order to assess the computational complexity, the procedure presented in [48] has been conducted. Under this process, C is obtained through the subsequent method:

1. The time reference T_0 is computed. T_0 represents the computing time consumed by the execution of the following standard code:

```
for j = 1:1000000
v = 0.55 + j
v = v + v; v = v/2;
v = v*v;
v = sqrt(v);
v = exp(v);
v = v/(v + 2);
end
```

2. Evaluate the computing time T_1 for function operation. T_1 expresses the time in which 200,000 runs of function f_9 (multimodal) are executed (only the function without optimization method). In the test, the function f_9 is operated with $n = 100$.

3. Calculate the execution time T_2 for the optimization algorithm. T_2 exhibits the elapsed time in which 200000 function evaluations of f_9 are executed (here, optimization method and function are combined).

4. The average time $\overline{T}_2$ is computed. First, execute the Step 3 five times. Then, extract their average value $\overline{T}_2 = (T_2^1 + T_2^2 + T_2^3 + T_2^4 + T_2^5)/5$.

5. The computational complexity C is obtained as follows: $C = (\overline{T}_2 - T_1)/T_0$.

Under this process, the computational complexity (C) values of HS, BAT, DE, PSO, CMA-ES, ABC, and FUZZY are obtained. Their values correspond to 77.23, 81.51, 51.20, 36.87, 40.77, 70.17 and 40.91, respectively. A smaller C value indicates that the method is less complex, which allows a faster execution speed under the same evaluation conditions. An analysis of the experiment results shows that although the proposed FUZZY algorithm is slightly more complex than PSO and CMA-ES, their computational complexity(C) values are comparable. Additionally, the three algorithms are significantly less computationally complex than HS, BAT, DE and ABC.

3.6 Conclusions

New methods in the area of metaheuristic algorithms have been able to demonstrate interesting performances in solving optimization problems. Greatest of them use operators based on metaphors of natural or social elements to evolve candidate solutions. Although humans have manifested their potential to resolve real-life complex optimization problems, the use of human intelligence to build optimization algorithms has been small widespread than the natural or social metaphors. A methodology to implement human-knowledge-based optimization strategies has been presented. Below the approach, a conducted search tactics are modeled in the rule base of a Takagi-Sugeno Fuzzy inference system, so that the implemented fuzzy rules express the conditions under which candidate solutions are evolved during the optimization process. All the approaches reported in the literature that integrate Fuzzy logic and metaheuristic techniques consider the optimization capabilities of the metaheuristic algorithms for improving the performance of fuzzy systems. In the proposed method, the approach is completely different. Under this new schema, the Fuzzy system direct conducts the search strategy during the optimization process. In this chapter, our intent is to propose a methodology for emulating human search strategies in an algorithmic structure. To the best of our knowledge, this is the first time that a fuzzy system is used as a metaheuristic algorithm. The proposed methodology presents three important characteristics: (1) Generation. Under the proposed methodology, fuzzy logic provides a simple and well-known method for constructing a search strategy via the use of human knowledge. (2) Transparency. It generates fully interpretable models whose content expresses the search strategy as humans can conduct it. (3) Improvement. AS human experts interact with an optimization process, they obtain a better comprehension of successful search strategies able to find optimal solutions. As a result, new rules are added so that their inclusion in the existing rule base improves the quality of the original search strategy. Under the proposed methodology, new rules can be easily incorporated into an already existent system. The addition of such rules allows the capacities of the original system to be extended. The proposed fuzzy algorithm has been experimentally evaluated with a test suite of 19 benchmark functions to confirm the efficacy and robustness of our method. The method has been compared to other popular optimization approaches based on evolutionary principles currently in employment. The results, statistically validated, have demonstrated that the proposed algorithm outperforms its competitors for most of the test functions in terms of its solution quality and convergence.

Appendix A. List of Benchmark Functions

Tables 3.17, 3.18, 3.19.

Table 3.17 Unimodal test functions used in the experimental study

Function	S	Dim	Minimum
$f_1(x) = \sum\limits_{i=1}^{n} x_i^2$	$[-100, 100]^n$	$n = 50$ $n = 100$	$x^* = (0, \ldots, 0);$ $f(x^*) = 0$
$f_2(x) = \sum\limits_{i=1}^{n} |x_i| + \prod\limits_{i=1}^{n} |x_i|$	$[-10, 10]^n$	$n = 50$ $n = 100$	$x^* = (0, \ldots, 0);$ $f(x^*) = 0$
$f_3(x) = \sum\limits_{i=1}^{n} \left(\sum\limits_{j=1}^{i} x_j \right)^2$	$[-100, 100]^n$	$n = 50$ $n = 100$	$x^* = (0, \ldots, 0);$ $f(x^*) = 0$
$f_4(x) = \max\limits_{i} \{|x_i|, 1 \leq i \leq n\}$	$[-100, 100]^n$	$n = 50$ $n = 100$	$x^* = (0, \ldots, 0);$ $f(x^*) = 0$
$f_5(x) = \sum\limits_{i=1}^{n-1} \left[100(x_{i+1} - x_i^2) + (x_i - 1)^2 \right]$	$[-30, 30]^n$	$n = 50$ $n = 100$	$x^* = (0, \ldots, 0);$ $f(x^*) = 0$
$f_6(x) = \sum\limits_{i=1}^{n} (x_i + 0.5)^2$	$[-100, 100]^n$	$n = 50$ $n = 100$	$x^* = (0, \ldots, 0);$ $f(x^*) = 0$
$f_7(x) = \sum\limits_{i=1}^{n} i x_i^4 + random(0, 1)$	$[-1.28, 1.28]^n$	$n = 50$ $n = 100$	$x^* = (0, \ldots, 0);$ $f(x^*) = 0$

Table 3.18 Multimodal test functions used in the experimental study

Function	S	Dim	Minimum
$f_8(x) = \sum_{i=1}^{n} x_i \sin\left(\sqrt{|x_i|}\right)$	$[-100, 100]^n$	$n = 50$ $n = 100$	$x^* =$ $(0, \ldots, 0)$; $f(x^*) = 0$
$f_9(x) = \sum_{i=1}^{n} \left[x_i^2 - 10\cos(2\pi x_i) + 10\right]$	$[-10, 10]^n$	$n = 50$ $n = 100$	$x^* =$ $(0, \ldots, 0)$; $f(x^*) = 0$
$f_{10}(x) = -20exp\left(-0.2\sqrt{\frac{1}{n}\sum_{i=1}^{n} x_i^2}\right)$ $- exp\left(\frac{1}{n}\sum_{i=1}^{n}\cos(2\pi x_i)\right) + 20 + exp$	$[-100, 100]^n$	$n = 50$ $n = 100$	$x^* =$ $(0, \ldots, 0)$; $f(x^*) = 0$
$f_{11}(x) = \frac{1}{4000}\sum_{i=1}^{n} x_i^2 - \prod_{i=1}^{n}\cos\left(\frac{x_i}{\sqrt{i}}\right) + 1$	$[-100, 100]^n$	$n = 50$ $n = 100$	$x^* =$ $(0, \ldots, 0)$; $f(x^*) = 0$
$f_{12}(x) = \frac{\pi}{n}\{10\sin(\pi y_1)$ $+ \sum_{i=1}^{n-1}(y_1 - 1)^2\left[1 + 10sin^2(\pi y_{i+1})\right]$ $+ (y_n - 1)^2\}$ $+ \sum_{i=1}^{n} u(x_i, 5, 100, 4);$ $y_i = 1 + \frac{(x_i + 1)}{4}$ $u(x_i, a, k, m)$ $= \begin{cases} k(x_i - a)^m & x_i > a \\ 0 & -a \leq x_i \leq a \\ k(-x_i - a)^m & x_i < a \end{cases}$	$[-30, 30]^n$	$n = 50$ $n = 100$	$x^* =$ $(0, \ldots, 0)$; $f(x^*) = 0$
$f_{13}(x) = 0.1\{sin^2(3\pi x_i)$ $+ \sum_{i=1}^{n}(x_i - 1)^2\left[1 + sin^2(3\pi x_i + 1)\right]$ $+ (x_n - 1)^2\left[1 + sin^2(2\pi x_n)\right]\}$ $+ \sum_{i=1}^{n} u(x_i, 5, 100, 4);$ $u(x_i, a, k, m) = \begin{cases} k(x_i - a)^m & x_i > a \\ 0 & -a < x_i < a \\ k(-x_i - a)^m & x_i < -a \end{cases}$	$[-100, 100]^n$	$n = 50$ $n = 100$	$x^* =$ $(0, \ldots, 0)$; $f(x^*) = 0$

(continued)

Table 3.18 (continued)

Function	S	Dim	Minimum
$f_{14}(x) = \frac{1}{2} \sum_{i=1}^{n} \left(x_i^4 - 16x_i^2 + 5x_i \right)$	$[-1.28, 1.28]^n$	$n = 50$ $n = 100$	$x^* =$ $(0, \ldots, 0);$ $f(x^*) = 0$

Table 3.19 Hybrid test functions used in the experimental study

Function	S	Dim	Minimum
$f_{15}(x) = f_1(x) + f_2(x) + f_9(x)$	$[-100, 100]^n$	$n = 50$ $n = 100$	$x^* = (0, \ldots, 0);$ $f(x^*) = 0$
$f_{16}(x) = f_5(x) + f_9(x) + f_{11}(x)$	$[-100, 100]^n$	$n = 50$ $n = 100$	$x^* = (0, \ldots, 0);$ $f(x^*) = 0$
$f_{17}(x) = f_3(x) + f_5(x) + f_{10}(x) + f_{13}(x)$	$[-100, 100]^n$	$n = 50$ $n = 100$	$x^* = (0, \ldots, 0);$ $f(x^*) = 0$
$f_{18}(x) = f_2(x) + f_5(x) + f_9(x) + f_{10}(x) + f_{11}(x)$	$[-100, 100]^n$	$n = 50$ $n = 100$	$x^* = (0, \ldots, 0);$ $f(x^*) = 0$
$f_{19}(x) = f_1(x) + f_2(x) + f_8(x) + f_{10}(x) + f_{12}(x)$	$[-100, 100]^n$	$n = 50$ $n = 100$	$x^* = (0, \ldots, 0);$ $f(x^*) = 0$

References

1. Zadeh, L.A.: Fuzzy sets. Inf. Control **8**, 338–353 (1965)
2. He, Yingdong, Chen, Huayou, He, Zhen, Zhou, Ligang: Multi-attribute decision making based on neutral averaging operators for intuitionistic fuzzy information. Appl. Soft Comput. **27**, 64–76 (2015)
3. Taur, J., Tao, C.W.: Design and analysis of region-wise linear fuzzy controllers, IEEE Trans. Syst. Man Cybern. Part B: Cybern. **27**(3), 526–532 (1997)
4. Ali, M.I., Shabir, M.: Logic connectives for soft sets and fuzzy soft sets. IEEE Trans. Fuzzy Syst. **22**(6), 1431–1442 (2014)
5. Novák, V., Hurtík, P., Habiballa, H., Štepnička, M.: Recognition of damaged letters based on mathematical fuzzy logic analysis. J. Appl. Logic **13**(2), 94–104
6. Papakostas, G.A., Hatzimichailidis, A.G., Kaburlasos, V.G.: Distance and similarity measures between intuitionistic fuzzy sets: a comparative analysis from a pattern recognition point of view. Pattern Recognit. Lett. **34**(14), 1609–1622 (2013)
7. Wang, Xinyu, Mengyin, Fu, Ma, Hongbin, Yang, Yi: Lateral control of autonomous vehicles based on fuzzy logic. Control Eng. Pract. **34**, 1–17 (2015)
8. Castillo, Oscar, Melin, Patricia: A review on interval type-2 fuzzy logic applications in intelligent control. Inf. Sci. **279**, 615–631 (2014)
9. Raju, G., Nair, M.S.: A fast and efficient color image enhancement method based on fuzzy-logic and histogram, AEU Int. J. Electron. Commun. **68**(3), 237–243 (2014)

10. Zareiforoush, H., Minaei, S., Alizadeh, M.R., Reza, M., Banakar, A.: A hybrid intelligent approach based on computer vision and fuzzy logic for quality measurement of milled rice. Measurement **66**, 26–34 (2015)
11. Nanda, S.J., Panda, G.: A survey on nature inspired metaheuristic algorithms for partitional clustering. Swarm Evol. Comput. **16**, 1–18 (2014)
12. Kennedy, J., Eberhart, R.: Particle swarm optimization. In: Proceedings of the 1995 IEEE International Conference on Neural Networks vol. 4, 1942–1948 (1995)
13. Karaboga, D.: An Idea Based on Honey Bee Swarm for Numerical Optimization. Technical Report-TR06. Engineering Faculty, Computer Engineering Department, Erciyes University (2005)
14. Geem, Z.W., Kim, J.H., Loganathan, G.V.: A new heuristic optimization algorithm: harmony search. Simulations **76**, 60–68 (2001)
15. Yang, X.S.: A new metaheuristic bat-inspired algorithm. In: Cruz, C., González, J., Krasnogor, G.T.N., Pelta, D.A. (eds.) Nature Inspired Cooperative Strategies for Optimization (NISCO 2010), Studies in Computational Intelligence, vol. 284, pp. 65–74. Springer, Verlag Berlin (2010)
16. Yang, X.S.: Firefly algorithms for multimodal optimization, in: stochastic algorithms: foundations and applications, SAGA 2009, SAGA Lect. Notes Comput. Sci. **5792** 169–178 (2009)
17. Cuevas, E., Cienfuegos, M., Zaldívar, D., Cisneros, M.P.: A swarm optimization algorithm inspired in the behavior of the social spider. Expert Syst. Appl. **40**(16), 6374–6384 (2013)
18. Erik, C., Mauricio, G., Daniel, Z., Marco, P.-C., Guillermo, G.: An algorithm for global optimization inspired by collective animal behaviour, Discrete Dyn. Nat. Soc. 2012 (2012) 24, http://dx.doi.org/10.1155/2012/638275. (Article ID 638275)
19. Storn, R., Price, K.: Differential Evolution—A Simple and Efficient Adaptive Scheme for Global Optimisation over Continuous Spaces, Technical ReportTR-95-012. ICSI, Berkeley, CA (1995)
20. Goldberg, D.E.: Genetic Algorithm in Search Optimization and Machine Learning, Addison-Wesley (1989)
21. Herrera, Francisco: Genetic fuzzy systems: taxonomy, current research trends and prospects. Evol. Intel. **1**, 27–46 (2008)
22. Alberto, F., Victoria, L., María José del, J., Francisco, H.: Revisiting evolutionary fuzzy systems: taxonomy, applications, new trends and challenges, Knowl.-Based Syst. **80** 109–121 (2015)
23. Caraveo, C., Valdez, F., Castillo, O.: Optimization of fuzzy controller design using a new bee colony algorithm with fuzzy dynamic parameter adaptation. Appl. Soft Comput. **43**, 131–142 (2016)
24. Castillo, O., Neyoy, H., Soria, J., Melin, P., Valdez, F.: A new approach for dynamic fuzzy logic parameter tuning in Ant Colony Optimization and its application in fuzzy control of a mobile robot. Appl. Soft Comput. **28**, 150–159 (2015)
25. Olivas, F., Valdez, Fe, Castillo, O., Melin, P.: Dynamic parameter adaptation in particle swarm optimization using interval type-2 fuzzy logic. Soft. Comput. **20**(3), 1057–1070 (2016)
26. Castillo, O., Ochoa, P., Soria, J.: Differential evolution with fuzzy logic for dynamic adaptation of parameters in mathematical function optimization, Imprecis. Uncertainty Inf. Represent. Process 361–374 (2016)
27. Guerrero, M., Castillo, O., Valdez, M.G.: Fuzzy dynamic parameters adaptation in the Cuckoo Search Algorithm using Fuzzy logic, CEC 441–448 (2015)
28. Alcala, R., Gacto, M.J., Herrera, F.: A fast and scalable multi-objective genetic fuzzy system for linguistic fuzzy modeling in high-dimensional regression problems. IEEE Trans. Fuzzy Syst. **19**(4), 666–681 (2011)
29. Alcala Fdez, J., Alcala, R., Gacto, M.J., Herrera, F.: Learning the membership function contexts for mining fuzzy association rules by using genetic algorithms, Fuzzy Sets Syst. **160**(7) 905–921 (2009)
30. Alcala, R., Alcala Fdez, J., Herrera, F.: A proposal for the genetic lateral tuning of linguistic fuzzy systems and its interaction with rule selection, IEEE Trans. Fuzzy Syst. **15**(4) 616–635 (2007)

31. Alcala Fdez, J., Alcala, R., Herrera, F.: A fuzzy association rule based classification model for high-dimensional problems with genetic rule selection and lateral tuning, IEEE Trans. Fuzzy Syst. **19**(5) 857–872 (2011)
32. Carmona, C.J., Gonzalez, P., del Jesus, M.J., Navio Acosta, M., Jimenez Trevino, L.: Evolutionary fuzzy rule extraction for subgroup discovery in a psychiatric emergency department, Soft Comput. **15**(12) 2435–2448 (2011)
33. Cordon, O.: A historical review of evolutionary learning methods for Mamdani type fuzzy rule-based systems: designing interpretable genetic fuzzy systems. Int. J. Approx. Reason. **52**(6), 894–913 (2011)
34. Cruz-Ramirez, M., Hervas Martinez, C., Sanchez Monedero, J., Gutierrez, P.A.: Metrics to guide a multi-objective evolutionary algorithm for ordinal classification. Neurocomputing 135 21–31 (2014)
35. Lessmann, S., Caserta, M., Arango, I.D.: Tuning metaheuristics: a data mining based approach for particle swarm optimization. Expert Syst. Appl. **38**(10), 12826–12838 (2011)
36. Sörensen, K.: Metaheuristics the metaphor exposed. Int. Trans. Oper. Res. **22**(1), 3–18 (2015)
37. Omid, M., Lashgari, M., Mobli, H., Alimardani, R., Mohtasebi, S., Hesamifard, R.: Design of fuzzy logic control system incorporating human expert knowledge for combine harvester. Expert Syst. Appl. **37**(10), 7080–7085 (2010)
38. Fullér, R., MJC, L.C., Darós, L.C.: Transparent fuzzy logic based methods for some human resource problems. Revista Electrónica de Comunicaciones y Trabajos de ASEPUMA **13**, 27–41 (2012)
39. Cordón, O., Herrera, F.: A three-stage evolutionary process for learning descriptive and approximate fuzzy-logic-controller knowledge bases from examples. Int. J. of Approx. Reason. **17**(4), 369–407 (1997)
40. Takagi, T., Sugeno, M.: Fuzzy identification of systems and its applications to modeling and control, IEEE Trans. Syst. Man Cybern. SMC-15 116–132 (1985)
41. Mamdani, E., Assilian, S.: An experiment in linguistic synthesis with a fuzzy logic controller. Int. J. Man Mach. Stud. **7**, 1–13 (1975)
42. Bagis, Aytekin, Konar, Mehmet: Comparison of Sugeno and Mamdani fuzzy models optimized by artificial bee colony algorithm for nonlinear system modelling. Trans. Inst. Meas. Control **38**(5), 579–592 (2016)
43. Guney, K., Sarikaya, N.: Comparison of mamdani and sugeno fuzzy inference system models for resonant frequency calculation of rectangular microstrip antennas. Prog. Electromagn. Res. B **12**, 81–104 (2009)
44. Baldick, R.: Applied Optimization. Cambridge University Press (2006)
45. Dan, S.: Evolutionary Algorithms Biologically Inspired and Population Based Approaches to Computer Intelligence. Wiley (2013)
46. Wong, S.Y., Yap, K.S., Yap, H.J., Tan, S.C., Chang, S.W.: On equivalence of FIS and ELM for interpretable rule-based knowledge representation. IEEE Trans. Neural Netw. Learn. Syst. **26**(7), 1417–1430 (2014)
47. Keem, S.Y., Shen, Y.W., Sieh, K.T: Compressing and improving fuzzy rules using genetic algorithm and its application to fault detection, IEEE 18th Conference on Emerging Technologies & Factory Automation (ETFA) vol. 1 1–4 (2013)
48. Liang, J.J., Qu, B.-Y., Suganthan, P.N.: Problem definitions and evaluation criteria for the CEC 2015 special session and competition on single objective real parameter numerical optimization. Technical Report 2013 11. Computational Intelligence Laboratory, Zhengzhou University, Zhengzhou China and Nanyang Technological University, Singapore (2015)
49. Nikolaus, H., Andreas, O., Andreas, G: On the adaptation of arbitrary normal mutation distributions in evolution strategies: the generating set adaptation, In: Proceedings of the 6th International Conference on Genetic Algorithms 57–64 (1995)
50. Boussaïda, I., Lepagnot, J., Patrick, S.: A survey on optimization metaheuristics. Inf. Sci. **237**, 82–117 (2013)
51. James, J., Yu, Q., Victor O., Li, K.: A social spider algorithm for global optimization. Appl. Soft Comput. **30**, 614–627 (2015)

52. Li, MD., Zhao, H., Weng, X.W., Han, T.: A novel nature-inspired algorithm for optimization: virus colony search. Adv. Eng. Softw. **92**, 65–88 (2016)
53. Han, M., Liu, C., Xing, J.: An evolutionary membrane algorithm for global numerical optimization problems. Inf. Sci. **276**, 219–241 (2014)
54. Meng, Z., Jeng, S.P.: Monkey King Evolution: a new memetic evolutionary algorithm and its application in vehicle fuel consumption optimization, Knowl.-Based Syst. **97** 144–157 (2016)
55. http://www.lri.fr/%E2%88%BChansen/cmaesintro.html
56. Wilcoxon, F.: Individual comparisons by ranking methods. Biometrics **1**, 80–83 (1945)
57. Garcia, S., Molina, D., Lozano, M., Herrera, F.: A study on the use of nonparametric tests for analyzing the evolutionary algorithms' behavior: a case study on the CEC'2005 Special session on real parameter optimization. J. Heuristics (2008). https://doi.org/10.1007/s10732-008-9080-4
58. Cuevas, E.: Block-matching algorithm based on harmony search optimization for motion estimation. Appl. Intell. **39**(1), 165–183 (2913)
59. Díaz-Cortés, M.-A., Ortega-Sánchez, N., Hinojosa, S., Cuevas, E., Rojas, R., Demin, A.: A multi-level thresholding method for breast thermograms analysis using Dragonfly algorithm. Infrared Phys. Technol. **93**, 346–361 (2018)
60. Díaz, P., Pérez-Cisneros, M., Cuevas, E., Hinojosa, S., Zaldivar, D.: An im-proved crow search algorithm applied to energy problems. Energies **11**(3), 571 (2018)

Chapter 4
A Metaheuristic Computation Scheme to Solve Energy Problems

Abstract The development of methods to solve optimization problems has increased in recent years. However, many of these methods are not tested in real applications. Energy problems are a topic of high relevance within the scientific community due to its environmental repercussions. Two representative cases involved in the huge energy consumption are the capacitor placement in radial distribution networks and parameter identification in induction motors. Both problems have intrinsically complex characteristics from the optimization perspective, which makes it difficult to solve them by conventional optimization methods. Some of these properties are their high multi-modality, discontinuity and non-linearity. Alternatively, meta-heuristic techniques have had shown performance in solving the solution to a wide range of complex engineering problems. A recent metaheuristic based on the intelligent group behavior of crows and their interaction is the Crow Search Algorithm (CSA). Although CSA shows interesting aspects, its performance in exploration mechanism exhibits considerable disadvantages when it confronts high multi-modal conditions. In this chapter, an enhance variant of the CSA approach is introduced to face complex optimization formulations of energy. The improved method is focused on modifying two main features from the original CSA: (I) the random perturbation and (II) the fixed awareness probability (AP). With the modifications, the enhance methodology conserves diversity in global solution and enhances the convergence to difficult high multi-modal optima. The performance evaluation of the presented methodology is addressed in a series of optimization problems related to distribution networks and inductions motors. The results of the evaluation show the robust performance of the presented methodology when it is contrasted with popular approaches in the literature.

4.1 Introduction

Energy problems are a topic of high relevance within the scientific community due to its environmental repercussions. Two representative cases involved in the huge

© Springer Nature Switzerland AG 2021 91
E. Cuevas et al., *Metaheuristic Computation: A Performance Perspective*,
Intelligent Systems Reference Library 195,
https://doi.org/10.1007/978-3-030-58100-8_4

energy consumption are the capacitor placement in radial distribution networks and parameter identification in induction motors.

Induction motors are an essential component in the industry field bringing several benefits, such as low cost, cheap maintenance, ruggedness, and easy to operate. These advantages have made induction motors widely used in industry, consuming up 2/3 of industrial energy [1, 2]. Due to the significant impact in the energy use has arisen the necessity to enhance their effectiveness, that is very correlated to the configuration of their internal parameters. The procedure to identify the internal parameters is an engineering challenge, mainly by their non-linearity and complexity nature. In this sense, the identification parameters problem in induction motors is an active research area. As a result, many approaches have been proposed to solve it in literature [3, 4].

On the other hand, distribution networks correspond to an open study area in electrical system. The distribution system, in conjunction with transmission, and generators are the three essentials components of a power system. In particular, 13% of the energy losses from the total generated are the responsibility of the electric distribution networks [5, 6]. This amount of energy loss present in distribution systems is mainly related to the scarcity of reactive power in the buses. An alternative to reduce such energy losses is compensating the reactive power in the distribution network through the allocation of capacitor banks. However, capacitor banks placement in a distribution network is a challenging problem that can be expressed as a combinatorial optimization problem, where the number and the sizes of capacitor banks, as well as a favorable position, must be optimally configured filling the network requirements. To achieve these goals and solve the combinatorial optimization problem different approaches have been introduced, which can be classified into four main categories; heuristic [11–13], analytical [5, 7, 8], based on artificial intelligence [14, 15], and numerical [9, 10]. A broad and detailed study of these methods is given in [15–17].

The high multimodality, discontinuity, and nonlinearity are characteristics that, from the point of view of optimization, make the problems of capacitor allocation in distribution networks and parameter identification in induction motors extremely complex to solve by traditional optimization methods.

The development of methods to solve optimization problems has increased in recent years. The metaheuristics techniques emerged as an alternative to the classic optimization methods and have gained great popularity due to their robustness to solve a wide variety of complex problems with high multi-modality, non-linearity, and discontinuity. The parameter identification of induction motors and the capacitor allocation problems have intrinsically complex characteristics that can be formulated as an optimization problem. Some works have addressed the problems with metaheuristic perspective such as the gravitational search algorithm [18, 19], bacterial foraging [20, 21], crow search algorithm [14], particle swarm optimization [22–25], genetic algorithm [26, 27], differential evolution [28–30], tabu search [31] and firefly [32].

A modern metaheuristic algorithm based on the emulating the intelligent behavior of groups of crows ant their interactions is the Crow Search Algorithm (CSA) [33]. Its simplicity and ease of implementation have attracted the attention of researches, and it has been applied in solving several complex engineering optimization problems such as water resources [35] and image processing [34], showing promising results. Despite its interesting results, the methodology suffers from certain difficulties in the search process when the objective function presents high multi-modality.

Due to difficulties in the original CSA search strategy when dealing with high multi-modal problems, this chapter proposes an improved variant of the CSA algorithm to solve high multi-modal problems. This chapter addressed two complex issues related to these characteristics: the identification of parameters in induction motors and the assignment of capacitors in distribution networks. The new version, named Improved Crow Search Algorithm (ICSA), modifies two fundamental elements to the original CSA: (I) the awareness probability (AP) parameter and (II) the random changes. In the first modification, the diversification-intensification ratio is improved by incorporating the dynamic awareness probability (DAP) parameter, which substitutes the fixed awareness probability (AP) parameter (I). The DAP is an adaptive parameter that is changing with the value of the cost function of each candidate solution. The ICSA adds the Lévy flight randomness to enhance the search capabilities of the traditional random perturbation (II) of CSA. The changes presented in the proposal allow ICSA to achieve the global optimum in complicated cost functions with a high multi-modal level, as well as to maintain the search capabilities of the algorithm. The performance evaluation of the presented methodology is addressed in a series of complex optimization problems related to distribution networks and inductions motors. Two models of induction motors integrate the first set of problems evaluated by the proposed algorithm. The second set of problems are four distribution networks of 10, 33, and 64 buses for optimal capacitor allocation. The experimental results are compared with similar state-of-the-art works and are also statistically analyzed.

The structure of the chapter is composed of the following sections: Sect. 4.2 defines the standar CSA, Sect. 4.3 the proposed methodology ICSA is described, Sect. 4.4 the internal parameter estimation problem of induction motor is presented, Sect. 4.5 the capacitor allocation problem is defined, and Sect. 4.6 the experimental results are exposed. Lastly, the conclusions are expressed in Sect. 4.7.

4.2 Crow Search Algorithm (CSA)

Metaheuristic schemes have demonstrated their interesting properties to solve complex engineering problems [53–55]. CSA is a representative metaheuristic method that has attracted the attention of evolutionary computation community. In the present section outlines the main concepts of the traditional CSA. Askarzadeh [33] developed a novel nature inspired metaheuristic methodology that attempts to emulate the smart conduct of crows. Crows are a very particular animal, their behavior presents a certain level of intelligence, which makes it a fascinating species. This intelligent behavior has shown the crow's ability to do things like utilizing tools, alert other crows to a potential enemy, identify faces, self-awareness, sophisticated communication techniques, and the ability to remember after a while places where they hide their food. Along with these behaviors, crows have a brain-body ratio very similar to the size of the human brain, which has contributed to crows being considered one of the most intelligent bird species in nature [36, 37].

The computational process of the CSA attempts to mimic the behavior of the crows in hiding the extra food and then retrieving it when necessary. The CSA is a population-based algorithm, where the length of the flock is composed of N candidates (crows) whose dimensionality is n, being n the dimension of the problem. The point in the search space $X_{i,k}$ of the crow i in a particular iteration k is expressed in Eq. (4.1) and describes a potential solution for the problem:

$$X_{i,k} = \left[x_{i,k}^1, x_{i,k}^2, \ldots, x_{i,k}^n\right] \quad i = 1, 2, \ldots, N \quad k = 1, 2, \ldots, \max Iter \qquad (4.1)$$

where max $Iter$ is the maximum number of repetitions in the method. It is assumed that each crow (individual) in the flock has the ability to memorize the best place visited $M_{i,k}$ until the current iteration to store food Eq. (4.2):

$$M_{i,k} = \left[m_{i,k}^1, m_{i,k}^2, \ldots, m_{i,k}^n\right] \qquad (4.2)$$

The CSA has two mechanisms to modify the position of each crow: persecution and escapement.

Persecution: In the pursuit stage, two crows are selected, a crow i as a thief and a crow j as a victim. The thief crow aims to find the unknown place of the victim crow. The thief crow aims to find the hidden place of the victim crow. In this step, the crow j does not perceive the proximity of the thief, and the thief achieves its goal.

Escapement: In the escapement stage, two crows take on the same role as the persecution stage, with the difference that the crow j notices the presence of the thief (crow i) and takes a stochastic flight to protect its food. This stochastic flight is computed in CSA by employing a random position change.

The type of movement of a certain crow i in iteration k is given by the awareness probability (AP) parameter of crow j, and a random value r_i between 0 and 1 uniformly distributed. Therefore, if r_i is less than AP parameter, the escapement movement is completed. Contrarily, persecution condition is selected. The model of this procedure is described as follows:

$$X_{i,k+1} = \begin{cases} X_{i,k} + r_i \cdot fl_{i,k} \cdot (M_{j,k} - X_{i,k}) & r_i \geq AP \\ random & otherwise \end{cases} \qquad (4.3)$$

The rate of change from crow $X_{i,k}$ to the fittest location $M_{j,k}$ of crow j is defined by the flight length $fl_{i,k}$ parameter.

After applying the movement operators of Eq. (4.3) to the flock, their new position is compared with their memory vector, and it is upgraded as illustrated below:

$$M_{i,k+1} = \begin{cases} F(X_{i,k+1}) & F(X_{i,k+1}) < F(M_{i,k}) \\ M_{i,k} & otherwise \end{cases} \qquad (4.4)$$

where the $F(\cdot)$ denotes the cost function to be optimized.

4.3 The Proposed Improved Crow Search Algorithm (ICSA)

Traditional CSA has shown competitive results finding optimal results for a set of particular solution spaces forms [14, 33, 38]. Nevertheless, the CSA presents a weak operator for the exploratory process which does not ensure the convergence to a global optimum. Following that circumstance, the exploration process has remarkable problems when it handles functions with high multi-modality. In the traditional CSA, the responsibility for the exploration process, falls on two primary components: The parameter AP (awareness probability) and the stochastic change (escapement). The parameter AP is what mange the equilibrium among exploration and exploitation in the search process. The random value on its part influences directly to the diversification stage by relocating the crow to a new location in the search space. Because of the significant implications of the components AP and a random value for achieving the optimum in objective functions, the ICSA method proposes a modification for both the AP and the random value.

4.3.1 Dynamic Awareness Probability (DAP)

In the original CSA, the AP value belongs to the initialization parameters and prevails fixed from start to end in the optimization task. This condition of the AP parameter is not positive for the diversification-intensification balance. With the purpose to enhance this balance, a parameter called dynamic awareness probability (DAP) is proposed to replace the fixed parameter AP. The DAP is an adaptable parameter, this is, its value is changing according to the aptitude of every promising solution. The employment of probability parameters related to adaptability value to the objective function had been effectively employed in the evolutionary research [39]. The mathematical model for the dynamic awareness probability (DAP) is as follows:

$$DAP_{i,k} = 0.9 \cdot \frac{F(X_{i,k})}{wV} + 0.1 \qquad (4.5)$$

where wV is the value of the less suitable crow until generation k. Considering a minimization task, this number is obtained from the population with the following calculation $wV = \max(F(X_{i,k}))$. Based on this probabilistic perspective, the most suitable candidate solutions will have a greater possibility to be selected. On the other hand, candidates with lower adaptability values will have a greater chance of being relocated in the search space to a random position.

4.3.2 Random Movement—Lévy Flight

Traditional CSA movements are determined by emulating two specific conducts of the crows: persecution and escapement. The conduct of escapement is imitated through a random change that is quantified by a random value uniformly distributed.

In nature, the adoption of techniques to seek a source of food is critical to survival. A poor exploration technique for finding prominent food sources can represent a threat to the animal life. In 1937 Paul Lévy introduces a kind of random walk that has been viewed in nature as a foraging patron and it is known as Lévy flights [40–42]. In Lévy flights, the movement dimension is regulated by a long-tailed probability distribution commonly named as Lévy distribution. The efficiency of Lévy Flights is superior to the uniform random distribution in terms solution space exploration [43].

In order to have a better tool to explore the search space, in the presented ICSA, the uniform random changes are substituted by the Lévy flights to emulate the escapement behavior. Hence, to generate a new random location $X_{i,k+1}$ by means of a Lévy flight method, a calculated Lévy flight value L is added as a disturbance to the current location $X_{i,k}$.

Mantegna procedure [44] is utilized to get a stable and symmetric L value of Lévy distribution. The first phase on Mantegna methodology is the calculation of the step size Z_i as follows:

$$Z_i = \frac{a}{|b|^{1/\beta}} \tag{4.6}$$

where a and b are n-dimensional vectors and $\beta = 3/2$. The elements of each vector a and b are sampled from the normal distribution characterized by the following parameters:

$$a \sim N\left(0, \sigma_a^2\right) \qquad b \sim N\left(0, \sigma_b^2\right)$$
$$\sigma_a = \left\{ \frac{\Gamma(1+\beta)sin(\pi\beta/2)}{\Gamma[(1+\beta)/2]\beta 2^{(\beta-1)/2}} \right\}^{1/\beta}, \qquad \sigma_b = 1 \tag{4.7}$$

where $\Gamma(\cdot)$ denotes a Gamma distribution. After obtaining the value of Z_i, the factor L is calculated by the following model:

$$L = 0.01 - Z_i \odot \left(X_{i,k} - X^{best}\right) \tag{4.8}$$

where the product $\odot$ implies the element-wise multiplications, X^{best} represents the best solution seen so far in terms of the fitness quality. Finally, the new position $X_{i,k+1}$ is given by:

$$X_{i,k+1} = X_{i,k} + L \tag{4.9}$$

The proposed ICSA algorithm is given in the form of a flowchart in Fig. 4.1.

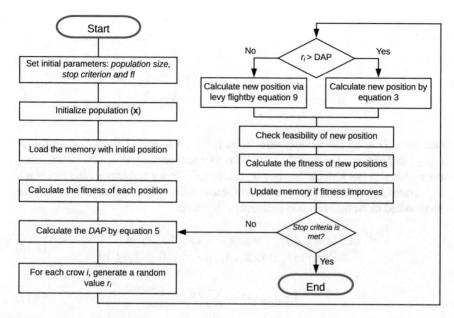

Fig. 4.1 Flowchart of ICSA algorithm

4.4 Motor Parameter Estimation Formulation

The procedure of extracting the internal parameter values directly on an induction motor is difficult due to the physical structure of its design. As an alternative to this circumstance, the parameters are estimated by means of identification techniques. The approximate and exact circuits are the two models that allow a proper estimation of internal parameters [4]. The number of parameters is the main difference between the two models.

The parameter estimation can be formulated as an optimization task whose dimension is given by the number of circuit model parameters. The main objective is to minimize the difference among the real manufacturing data and the estimated ones, adapting the equivalent circuit through the values of its parameters. Under these conditions, the procedure to identify the parameters in induction motors is a complex problem due to its high multimodality and solving it is a challenging task.

4.4.1 Approximate Circuit Model

The description of the motor using the approximate circuit model does not incorporate the rotor reactance and magnetization parameters. This makes the exact circuit model more accurate than the approximate circuit model.

The graphical description of the approximate circuit is shown in Fig. 4.2. The

Fig. 4.2 Approximate circuit model

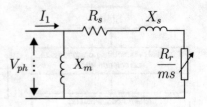

real data given by the manufacturer such full load torque (T_{fl}), maximum torque (T_{max}) and starting torque (T_{str}) are used to estimate the motor parameters, such the motor slip (ms), stator leakage reactance (X_s), rotor resistance (R_r) and stator resistance (R_s) shown in Fig. 4.2. Under these conditions, the estimation process can be modeled as an optimization problem, as follows:

$$\text{min Cost}_{AM}(\theta) \quad \text{where } \theta = (R_s, R_r, X_s, ms)$$
$$\text{Subject to} \quad 0 < R_s, R_r, ms < 1, 0 \le X_s \le 10 \tag{4.10}$$

$$\text{Cost}_{AM}(\theta) = \sum_{i=1}^{3} (F_i(\theta))^2 \tag{4.11}$$

$$F_1(\theta) = \frac{\frac{K_t R_r}{ms\left[\left(R_s + \frac{R_r}{ms}\right)^2 + X_s\right]} - T_{fl}}{T_{fl}} \tag{4.12}$$

$$F_2(\theta) = \frac{\frac{K_t R_r}{(R_s + R_r)^2 + X_s} - T_{str}}{T_{str}} \tag{4.13}$$

$$F_3(\theta) = \frac{\frac{K_t}{2\left[R_s + \sqrt{R_s^2 + X_s^2}\right]} - T_{max}}{T_{max}} \tag{4.14}$$

$$K_t = \frac{3V_{ph}^2}{\omega_s} \tag{4.15}$$

4.4.2 Exact Circuit Model

The exact circuit model contrasted with the approximate circuit model incorporates all the parameters involved in the description of the induction motor. The circuit diagram for the exact circuit model is illustrated in Fig. 4.3. This circuit incorporates two parameters, the magnetizing leakage reactance (X_m) and the rotor leakage reactance (X_s) to those considered by the approximate circuit, the motor slip (ms), stator leakage reactance (X_s), rotor resistance (R_r) and stator resistance (R_s), to achieve the full load torque (T_{fl}), maximum torque (T_{max}) and starting torque (T_{str}) and full

Fig. 4.3 Exact circuit model

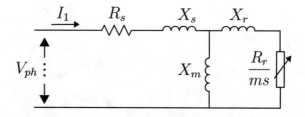

load power factor (pf). According to the characteristics of this exact circuit model, the parameter estimation process can be handled as an optimization problem in the following terms:

$$\min \text{Cost}_{EM}(\theta) \quad \text{where } \theta = (R_s, R_r, X_s, X_r, X_m, ms)$$
$$0 < R_s, R_r, X_s, X_r, ms < 1, 0 < X_m < 10 \quad (4.16)$$
$$\text{Subject to} \quad n_{fl} = \frac{p_{fl} - (I_1^2 R_s + I_1^2 R_r + P_{rot})}{p_{fl}}$$

where p_{fl} and P_{rot} are the rated power and rotational losses, while n_{fl} is the efficiency given by the manufacturer. The values of p_{fl} and P_{rot} are selected according [3, 18] to keep consistency with similar works:

$$\text{Cost}_{EM}(\theta) = \sum_{i=1}^{4} (F_i(\theta))^2 \quad (4.17)$$

$$F_1(\theta) = \frac{\frac{K_t R_r}{ms\left[\left(R_{th} + \frac{R_r}{ms}\right)^2 + X^2\right]} - T_{fl}}{T_{fl}} \quad (4.18)$$

$$F_1(\theta) = \frac{\frac{K_t R_r}{(R_{th} + R_r)^2 + X^2} - T_{str}}{T_{str}} \quad (4.19)$$

$$F_3(\theta) = \frac{\frac{K_t}{2\left[R_{th} + \sqrt{R_{th}^2 + X^2}\right]} - T_{max}}{T_{max} \times mf} \quad (4.20)$$

$$F_4(\theta) = \frac{\cos\left(\tan^{-1}\left(\frac{X}{R_{th} + \frac{R_r}{ms}}\right)\right) - pf}{pf} \quad (4.21)$$

$$V_{th} = \frac{V_{ph} X_m}{X_s + X_m} \quad (4.22)$$

$$R_{th} = \frac{R_s X_m}{X_s + X_m} \quad (4.23)$$

$$X_{th} = \frac{X_s X_m}{X_s + X_m} \tag{4.24}$$

$$K_t = \frac{3V_{ph}^2}{\omega_s} \tag{4.25}$$

$$X = X_r + X_{th} \tag{4.26}$$

4.5 Capacitor Allocation Problem Formulation

4.5.1 Load Flow Analysis

This section describes the capacitor allocation problem (CAP). Before considering a capacitor installation, it is convenient to perform a load flow analysis to know the current state of the distribution network in terms of power losses and voltage profile. Different methods to perform the load flow analysis have been proposed in literature [45, 46]. In this work, the approach proposed in [47] is used due to its simplicity to find out the voltage and power losses status of the network. Adopting a three-line balanced system as a single line diagram, as depicted in Fig. 4.4, the values calculation for power losses and voltage are conducted as follows:

$$|V_{i+1}| = -\left[\frac{V_i^2}{2} - R_i \cdot P_{i+1} - X_i \cdot Q_{i+1}\right] + \left[\left(-\frac{V_i^2}{2} + R_i \cdot P_{i+1} + X_i \cdot Q_{i+1}\right) - \left(R_i^2 + X_i^2\right) \cdot \left(P_i^2 + Q_i^2\right)\right]^{1/2} \tag{4.27}$$

$$P_{Li} = \frac{R_i \cdot [P_{i+1}^2 + Q_{i+1}^2]}{|V_{i+1}|^2} \tag{4.28}$$

$$Q_{Li} = \frac{X_i \cdot [P_{i+1}^2 + Q_{i+1}^2]}{|V_{i+1}|^2} \tag{4.29}$$

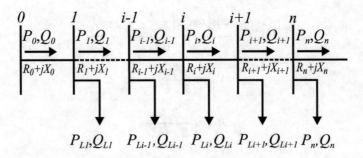

Fig. 4.4 Simple radial distribution system

$$P_{LOSS} = \sum_{i=1}^{N} P_{Li} \tag{4.30}$$

where R_i is the resistance and X_i the reactance of branch i, V_{i+1} is the voltage value in the $i - th + 1$ node. P_{i+1} is the real reactive power and Q_{i+1} correspond to the reactive power load flowing through node $i - th + 1$, while P_{LOSS} is the total real loss in the network.

4.5.2 Mathematical Approach

The OCA task over the buses of a distribution network can be formulated by the set of values that minimize the cost of installing the capacitors (Eq. 4.31) and the yearly cost caused by the real power loss of the entire system (Eq. 4.32):

$$Min \ IC = \sum_{i=1}^{N} k_i^c \cdot Q_i^c \tag{4.31}$$

$$Min \ AC = k_p \cdot P_{LOSS} \tag{4.32}$$

where IC is the installation cost of the capacitors, N represents the total of buses selected for a capacitor installation, k_i^c is the price per $kVar$, and Q_i^c stands for the size of capacitor in bus i, AC corresponds to the annual cost caused by the real power losses, k_p is taken as the annual price of the losses in kW, P_{LOSS} is the total of the real power losses in the system. Expenses generated by the capacitor maintenance is not considered in the objective function.

Considering the installation cost and the annual cost of the power losses, the final objective function is given as follows:

$$Min \ F = AC + IC \tag{4.33}$$

The objective function F (Eq. 4.33) is limited by voltage restrictions determined by:

$$V_{min} \leq |V_i| \leq V_{max} \tag{4.34}$$

where $|V_i|$ is the voltage magnitude in bus i, the lower and upper voltage boundaries are the $V_{min} = 0.9$ p.u. and $V_{max} = 1.0$ p.u. respectively.

Considering such aspects, the size, ubication, and the number of capacitors are the key elements to be optimally selected into the optimization process. The high dimensionality and the high multimodality make the OCA a complex optimization process to solve by traditional methods; for this purpose, this work propose the ICSA to solve it.

4.5.3 Sensitivity Analysis and Loss Sensitivity Factor

The sensitivity analysis (SA) is a method that its main goal is to reduce the search space. The SA gives information about the parameters to lessen the optimization task. In the OCA problem, the SE is applied to obtain the set of parameters with the least variability [48]. This knowledge permits to determine the best nodes that can be seen as a possible location for capacitor assignment. With such perspective, the exploration space can be decreased. In addition, the nodes that will have higher reduction in real power loss when installing the capacitors, will be those that SA method identified with les variability.

Assuming a simple distribution line from Fig. 4.4, as is shown in Fig. 4.5, the equations of active power loss and reactive power loss (Eqs. (4.28)–(4.29)) can be rewritten as follows:

$$P_{Lineloss}(i+1) = \frac{\left[P_{eff}^2(i+1) + Q_{eff}^2(i+1) \right] \cdot R(i)}{V^2(i+1)} \qquad (4.35)$$

$$Q_{Lineloss}(i+1) = \frac{\left(P_{eff}^2(i+1) + Q_{eff}^2(i+1) \right) \cdot X(i)}{V^2(i+1)} \qquad (4.36)$$

where the term $P_{eff}(i+1)$ correspond to the total effective active power presented further the node $i+1$, and $Q_{eff}(i+1)$ is equivalent to the effective value of reactive power supplied further the node $i+1$.

Therefore, the loss sensitivity factors now can be obtained from Eqs. (4.35) and (4.36) as follows:

$$\frac{\partial P_{Lineloss}}{\partial Q_{eff}} = \frac{2 \cdot Q_{eff}(i+1) \cdot R(i)}{V^2(i+1)} \qquad (4.37)$$

Fig. 4.5 Simple distribution line

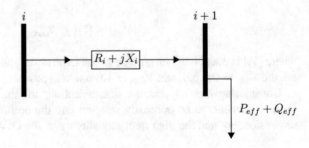

$$\frac{\partial Q_{Lineloss}}{\partial Q_{eff}} = \frac{2 \cdot Q_{eff}(i+1) \cdot X(i)}{V^2(i+1)} \tag{4.38}$$

Now, the processes to detect the possible candidate nodes is summarized in the next steps:

Step 1. Using Eq. (4.25) to compute the Loss Sensitivity Factors for all nodes.
Step 2. Sort in descending order the Loss Sensitivity Factors and its corresponding node index.
Step 3. Calculate the normalized voltage magnitudes for all nodes using:

$$norm(i) = \frac{V(i)}{0.95} \tag{4.39}$$

Step 4. Set a node as possible candidate those nodes whose norm value (calculated in the previous step) is less than 1.01.

4.6 Experiments

To test the performance of the proposed methodology, it is applied to a variety of experiments of two complex energy problems. The first set of experiments is the estimation of the internal values of two models of induction motor. The second collection of experiments aims to decrease the real power losses and improve the buses voltage in three different distribution networks by solving the OCA problem. The computer equipment used in the experiments is a Pentium dual-core with a 2.53 GHz CPU and 8-GB RAM.

4.6.1 Motor Parameter Estimation Test

The performance of the proposed ICSA is tested over two induction motors with the purpose to estimate their optimal parameters. In the experimental process, the approximate (CostAM) and exact (CostEM) circuit model are used. The manufacturer characteristics of the motors are presented in Table 4.1.

In the test, the results of the proposed ICSA method is compared to those presented by the popular algorithms DE, ABC and GSA. The parameters setting of the algorithms has been used in order to maintain compatibility with other works reported in the literature [18, 19, 28, 49] and are shown in Table 4.2.

The tuning parameter fl in the ICSA algorithm is selected as result of a sensitivity analysis which through experimentally way evidence the best parameter response. Table 4.3 shows the sensitivity analysis of the two energy problems treated in this work.

Table 4.1 Manufacturer's motor data

Characteristics	Motor 1	Motor 2
Power, HP	5	40
Voltage, V	400	400
Current, A	8	45
Frequency, Hz	50	50
No. Poles	4	4
Full load split	0.07	0.09
Starting torque (Tstr)	15	260
Max. Torque (Tmax)	42	370
Stator current	22	180
Full load torque (Tfl)	25	190

Table 4.2 Algorithms parameters

DE	ABC	GSA	ICSA
$CR = 0.5$	$\varphi_{ni} = 0.5$	$Go = 100$	$fl = 2.0$
$F = 0.2$	$SN = 120$	$Alpha = 20$	

Table 4.3 Sensitivity Analysis of fl parameter

fl value	Analysis	Motor 2 exact model	Motor 2 approximate model	10 bus	33 bus	69 bus
1.0	Min	7.1142×10^{-3}	1.4884×10^{-13}	696.6	139.9	145.9
	Max	1.1447×10^{-2}	2.6554×10^{-5}	699.6	146.1	164.1
	Mean	7.2753×10^{-3}	1.6165×10^{-5}	697.2	140.5	147.9
	Std	7.4248×10^{-4}	5.6961×10^{-5}	0.7	1.4	3.3
1.5	Min	7.1142×10^{-3}	0.0000	696.6	139.4	146.1
	Max	7.2485×10^{-3}	2.6172×10^{-3}	701.0	144.6	161.0
	Mean	7.1237×10^{-3}	7.47886×10^{-5}	697.7	140.7	147.5
	Std	2.5728×10^{-5}	4.4239×10^{-4}	0.8	1.9	2.8
2	Min	7.1142×10^{-3}	0.0000	696.6	139.2	145.8
	Max	7.1142×10^{-3}	1.1675×10^{-25}	698.1	140.3	156.4
	Mean	7.1142×10^{-3}	3.4339×10^{-27}	696.8	139.5	146.2
	Std	2.2919×10^{-11}	1.9726×10^{-26}	0.4	1.9	2.1
2.5	Min	7.1142×10^{-3}	4.3700×10^{-27}	696.8	140.4	145.9
	Max	7.1170×10^{-3}	2.3971×10^{-16}	698.8	148.8	158.7
	Mean	7.1142×10^{-3}	1.7077×10^{-17}	697.9	141.2	147.5
	Std	5.3411×10^{-7}	5.0152×10^{-17}	0.5	2.1	2.6
3	Min	7.1142×10^{-3}	6.8124×10^{-21}	696.8	140.5	146.0
	Max	7.1121×10^{-3}	2.9549×10^{-13}	703.4	158.7	164.1
	Mean	7.1142×10^{-3}	1.3936×10^{-14}	697.3	141.4	147.3
	Std	1.5398×10^{-7}	5.1283×10^{-14}	1.7	3.3	3.3

In the comparison, the algorithms are tested with a population size of 25 individuals, with a maximum number of generations established in 3000. This termination criterion as well as the parameter setting of algorithms has been used to maintain concordance with the literature [18, 28, 49]. Additionally, the results are analyzed and validated statistically through the Wilcoxon test.

The results for the approximate model ($Cost_{AM}$) produced by motor 1 and motor 2 are presented in Tables 4.4 and 4.5, respectively. In the case of exact model ($Cost_{EM}$) the values of standard deviation and mean for the algorithms are shown in Table 4.6 for motor 1 and the results for motor 2 in Table 4.7. The results presented are based on an analysis of 35 independent executions for each algorithm. The results demonstrate that the proposed ICSA method is better than its competitors in terms of accuracy (Mean) and robustness (Std.).

The comparison of the final fitness values of different approaches is not enough to validate a new proposal. Other additional test also represents the convergence graphs. They show the evolution of the solutions through the optimization process. Therefore, they indicate which approaches reach faster the optimal solutions. Figure 4.6 shows the convergence comparison between the algorithms in logarithmic scale for a better appreciation, being the proposed method, which present a faster to achieve the global optimal.

Table 4.4 Results of approximate circuit model CostAM, for motor 1

Analysis	DE	ABC	GSA	ICSA
Mean	1.5408×10^{-4}	0.0030	5.4439×10^{-21}	1.9404×10^{-30}
Std.	7.3369×10^{-4}	0.0024	4.1473×10^{-21}	1.0674×10^{-29}
Min	1.9687×10^{-15}	2.5701×10^{-5}	$3.4768 \times 10 - 22$	1.4024×10^{-32}
Max	0.0043	0.0126	1.6715×10^{-20}	6.3192×10^{-29}

Table 4.5 Results of approximate circuit model CostAM, for motor 2

Analysis	DE	ABC	GSA	ICSA
Mean	4.5700×10^{-4}	0.0078	5.3373×10^{-19}	3.4339×10^{-27}
Std.	0.0013	0.0055	3.8914×10^{-19}	1.9726×10^{-26}
Min	1.1369×10^{-13}	3.6127×10^{-4}	3.7189×10^{-20}	0.0000
Max	0.0067	0.0251	1.4020×10^{-18}	1.1675×10^{-25}

Table 4.6 Results of exact circuit model CostEM for motor 1

Analysis	DE	ABC	GSA	ICSA
Mean	0.0192	0.0231	0.0032	0.0019
Std.	0.0035	0.0103	0.0000	4.0313×10^{-16}
Min	0.0172	0.0172	0.0032	0.0019
Max	0.0288	0.0477	0.0032	0.0019

Table 4.7 Results of exact circuit model CostEM for motor 2

Analysis	DE	ABC	GSA	ICSA
Mean	0.0190	0.0791	0.0094	0.0071
Std.	0.0057	0.0572	0.0043	2.2919×10^{-11}
Min	0.0091	0.0180	0.0071	0.0071
Max	0.0305	0.2720	0.0209	0.0071

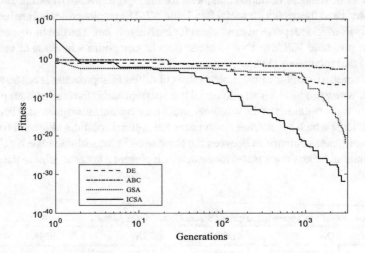

Fig. 4.6 Convergence evolution comparison of approximate model ($Cost_{AM}$) for motor 1

4.6.2   *Capacitor Allocation Test*

To test the potential of the proposed methodology, it is applied to solve the OCA problem in a three different distribution networks, the 10-bus [50], 33-bus [51], and 69-bus [52].

The experimental results are compared with the ABC, DE, GSA which are related algorithms. The tuning parameters used by the algorithms are shown in Table 4.2. All the experiments are done with 100 as maximum number of iterations and with 25 search individuals.

4.6.2.1   10-Bus System

The 10-bus distribution system with nine lines is considered as the first case. The single line diagram of the system is shown in Fig. 4.7, where the first bus is regarded as the substation bus. In the appendix section the Table 4.16 shows the system details, such as the real and reactive loads for the buses, as well as the resistance and reactance

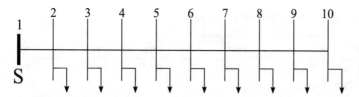

Fig. 4.7 10-bus distribution test system

on each line. The substation feeds the system with a 23 kV, while its load demands are a total of 12,368 kW and 4186 kVAr of real and reactive power respectively.

The network operating under normal conditions without any capacitor installed, presents a minimum and maximum voltage of 0.8404 and 0.9930 p.u. located in bus 10th and 2nd respectively, as well as a power loss of 783.77 kW. This work considers for all the experimental cases a power loss cost of $168 per kW, therefore for the 10-bus network whit a power loss of 783.77 kW corresponds a cost of $131,674. Sensitivity analysis is executed as the first stage of the proposal to identify the nodes that can potentially be chosen to allocate a capacitor. For the 10-bus system, the nodes 6, 5, 9, 10, 8 and 7 are the buses with higher probability to be chosen. The annual cost per kVAr and the capacitor sizes are taken considering a set of 27 standard values showing in Table 4.8. The detailed results obtained by ICSA through the optimization process are displayed in Table 4.9. After the ICSA performance the buses selected are 6, 5, 9, 10, 8 and their respective sizes of capacitor installed are 1200, 3900, 150, 600, 450 kVAr. The compensated system has a total power loss of 696.76 kW which generates a yearly cost of $117,055.68. The results achieve by the algorithms are compared in Table 4.9, while their convergence evolution are illustrated in Fig. 4.8.

Table 4.8 Possible capacitor sizes and cost ($/kVAr)

j	Q	$/kVAr	j	Q	$/kVAr	j	Q	$/kVAr
1	150	0.500	10	1500	0.201	19	2850	0.183
2	350	0.350	11	1650	0.193	20	3000	0.180
3	450	0.253	12	1800	0.187	21	3150	0.195
4	600	0.220	13	1950	0.211	22	3300	0.174
5	750	0.276	14	2100	0.176	23	3450	0.188
6	900	0.183	15	2250	0.197	24	3600	0.170
7	1050	0.228	16	2400	0.170	25	3750	0.183
8	1200	0.170	17	2550	0.189	26	3900	0.182
9	1350	0.207	18	2700	0.187	27	4050	0.179

Table 4.9 Experimental results of 10-bus system

Items algorithms	Base Case	Compensated			
		DE	ABC	GSA	ICSA
Total power losses (P_{Loss}), kW	783.77	700.24	697.18	699.67	696.76
Total power losses cost, $	131,673.36	117,640.32	117,126.24	117,544.56	117,055.68
Optimal buses	–	6, 5, 9, 10	6, 5, 10, 8	6, 5, 9, 10, 7	6, 5, 9, 10, 8
Optimal capacitor size	–	900, 4050, 600, 600	1200, 4050, 600, 600	1650, 3150, 600, 450, 150	1200, 3900, 150, 600, 450
Total kVAr	–	6150	6450	6000	6300
Capacitor cost, $	–	1153.65	1192.95	1253.55	1189.8
Total annual cost	131,673.36	118,793.97	118,329.19	118,798.11	118,245.48
Net saving, $	–	12,879.38	13,344.17	12,875.24	13,427.88
% saving		9.78	10.10	9.77	10.19
Minimum voltage, p.u.	0.8375 (bus 10)	0.9005 (bus 10)	0.9001 (bus 10)	0.9002 (bus 10)	0.9000 (bus 10)
Maximum voltage, p.u.	0.9929 (bus 2)	0.9995 (bus 3)	1.0001 (bus 3)	0.9992 (bus 3)	0.9997 (bus 3)

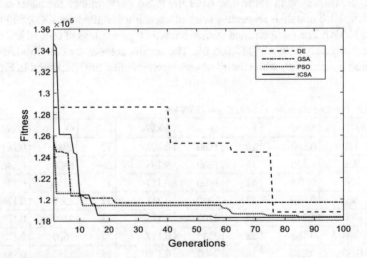

Fig. 4.8 Convergence comparison 10-bus

CSA Versus ICSA

In order to compare directly the original CSA version with the proposed ICSA, the same experiments conducted in [14] have been considered. In the first experiment, the optimization process involves only the candidate buses of 5, 6 and 10 for capacitor

Table 4.10 Experiments results of ICSA over CSA in 10-bus system

Items algorithms	Base Case	Experiment 1		Experiment 2	
		CSA	ICSA	CSA	ICSA
Total power losses (P_{Loss}), kW	783.77	698.14	698.14	676.02	675.78
Total power losses cost, $	131,673.36	117,287.52	117,287.52	113,571.36	113,531.04
Optimal buses	–	5, 6, 10	5, 6, 10	3, 4, 5, 6, 7, 10	3, 4, 5, 6, 8, 10
Optimal capacitor size	–	4050, 1650, 750	4050, 1650, 750	4050, 2100, 1950, 900, 450, 600	4050, 2400, 1650, 1200, 450, 450
Total kVAr	–	6150	6450	10,050	10,200
Capacitor cost, $	–	6450	6450	10,050	10,200
Total annual cost	–	118,537.92	118537.92	115,487.91	115,414.14
Net saving, $	131,673.36	13,135.43	13,135.43	16,185.44	16,259.21
% saving	–	9.9	9.9	12.29	12.35
Minimum voltage, p.u.	0.8375 (bus 10)	0.9000 (bus 10)	0.9000 (bus 10)	0.9003 (bus 10)	0.9000 (bus 10)
Maximum voltage, p.u.	0.9929 (bus 2)	1.0001 (bus 3)	1.0001 (bus 3)	1.0070 (bus 3)	1.0070 (bus 3)

allocation. The second test considers all the buses as possible candidates (except the substation bus) for capacitor installation. For the first experiment, all the possible capacitor combinations are $(27 + 1)3 = 21,952$. Under such conditions, it is possible to conduct a brute-force search for obtaining the global best. For this test, both algorithms (CSA and ICSA) have been able to achieve the global minimum.

In the second experiment, all buses are candidates for capacitor allocation. Under this approach, there are $(27 + 1)9 = 1.0578 \times 1013$ different combinations. In this scenario, a brute-force strategy is computationally expensive. The results of these experiments are shown in Table 4.10.

4.6.2.2 33-Bus System

In this experiment, a system with 33 buses and 32 lines is considered. In the system, the first bus is assumed as substation bus with a voltage of 12.66 kV. The network configuration is illustrated in Fig. 4.9.

The information about line resistance and reactance, as well as the corresponding load profile is shown in the Appendix in Table 4.17. The 33-bus distribution network before the capacitor installation presents a total power loss of 210.97 kW with an annual cost of $35,442.96 and a total active power of 3715 kW.

Once the optimization process is conducted, the buses 6, 30, 13 are selected as the optimal location with the corresponding sizes of 450, 900, 350 kVAr. The

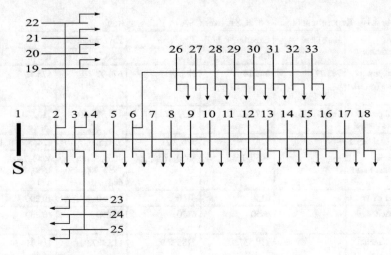

Fig. 4.9 33-bus distribution test system

candidate buses are determined by the sensitivity analysis. The total power loss after the capacitor installation is decreased from 210.91 to 138.74 kW, saving 32.59% of the original cost. The results of the test in detail and the comparison between the algorithms are shown in Table 4.11. Figure 4.10 illustrates the convergence evolution of the all algorithms.

CSA Versus ICSA

This section presents a direct comparison between original crow search algorithm (CSA) and the proposed method (ICSA). The 33-bus network is analyzed by CSA as is presented in [14] where the buses 11, 24, 30 and 33 are taken as candidates and the capacitor sizes and their kVAr values are shown in Table 4.12.

The results obtained from the both algorithms are compared in Table 4.13. The table shows that the ICSA is capable to obtain accurate results than the original version CSA.

4.6.2.3   69-Bus System

For the third capacitor allocation experiment a network of 68 buses and 69 lines is analyzed. Before to install any capacitor, the network presents a total active power loss of 225 kW, a total active and reactive power load of 3801.89 kW and 2693.60 kVAr. The annual cost for the corresponding 225 kW power loss is $37,800.00. The system presents a minimum voltage of 0.9091 p.u. at the 64th bus and a maximum 0.9999 p.u. at 2nd bus. As in the 10 and 33 bus experiments, the possible capacitor sizes and the price per kVAr is shown in the Table 4.8. The network diagram is

Table 4.11 Experimental results of 33-bus system

Items algorithms	Base Case	Compensated			
		DE	ABC	GSA	ICSA
Total power losses (P_{Loss}), kW	210.97	152.92	141.13	140.27	139.49
Total power losses cost, $	35,442.96	25,690.56	23,740.08	23,565.60	23,434.32
Optimal buses	–	6, 29, 30, 14	6, 29, 8, 13, 27, 31, 14	30, 26, 15	30, 7, 12, 15
Optimal capacitor size	–	350, 750, 350, 750	150, 150, 150, 150, 600, 450, 150	900, 450, 350	900, 600, 150, 150
Total kVAr	–	2200	1800	1700	1800
Capacitor cost, $	–	659	620.85	401.05	446.70
Total annual cost	35,442.96	26,349.56	25,540.00	23,966.65	23,881.02
Net saving, $	–	9093.39	9902.95	11,476.31	11,561.94
% saving	–	25.65	27.94	32.37	32.62
Minimum voltage, p.u.	0.9037 (bus 18)	0.9518 (bus 18)	0.9339 (bus 18)	0.9348 (bus 18)	0.9339 (bus 18)
Maximum voltage, p.u.	0.9970 (bus 2)	0.9977 (bus 2)	0.9976 (bus 2)	0.9975 (bus 2)	0.9976 (bus 2)

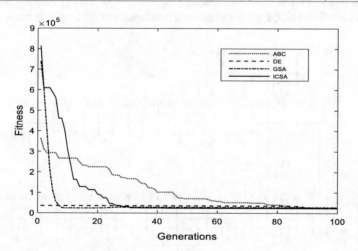

Fig. 4.10 Convergence comparison 33-bus

Table 4.12 Possible capacitor sizes and cost ($/kVAr)

j	1	2	3	4	5	6
Q	150	300	450	600	750	900
$/kVAr	0.500	0.350	0.253	0.220	0.276	0.183

Table 4.13 Experiment result of ICSA over CSA in 33-bus system

Items algorithms	Base Case	Compensated	
		CSA	ICSA
Total power losses (P_{Loss}), kW	210.97	139.30	138.14
Total Power losses cost, $	35,442.96	23,402.40	23,207.52
Optimal buses	–	11, 24, 30, 33	11, 24, 30, 33
Optimal capacitor size	–	600, 450, 600, 300	450, 450, 900, 150
Total kVAr	–	1950	1950
Capacitor cost, $	–	482.85	467.40
Total annual cost	35,442.96	23,885.25	23,674.92
Net saving, $	–	11,557.71	11,768.04
% saving	–	32.60	33.20
Minimum voltage, p.u.	0.9037 (bus 18)	0.9336 (bus 18)	0.9302 (bus 18)
Maximum voltage, p.u.	0.9970 (bus 2)	0.9976 (bus 2)	0.9976 (bus 2)

illustrated in Fig. 4.11 and the line and load data is presented in Table 4.18 in the Appendix.

Using the ICSA method, optimal buses selected are the 57, 61 and 18 with the capacitor values of 150, 1200, 350 kVAr respectively. With this reactance adjustment the total power loss is reduced from 225 to 146.20 kW, saving 33.96% from the original cost. The voltage profile presents a minimum of 0.9313 p.u. at bus 65th and

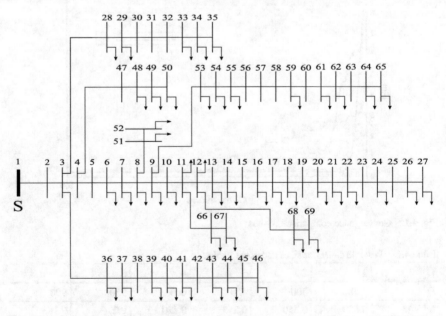

Fig. 4.11 69-bus distribution test system

Table 4.14 Experiment result of 69-bus system

Items algorithms	Base Case	compensated			
		DE	ABC	GSA	ICSA
Total power losses (P_{Loss}), kW	225	210.02	149.36	147.1017	146.20
Total power losses cost, $	37,800.00	35,283.36	25,092.48	24,712.80	24,561.60
Optimal buses	–	57, 58, 64, 21, 63, 20, 62, 26	58, 59, 62, 24	61, 27, 22	57, 61, 18
Optimal capacitor size	–	750, 600, 900, 350, 150, 150, 350, 150	150, 150, 900, 150	1200, 150, 150	150, 1200, 350
Total kVAr	–	3400	1350	1500	1700
Capacitor cost, $	–	973.70	389.70	354.00	401.50
Total annual cost	37800.00	36257.06	25,482.18	25,066.8	24,961.10
Net saving, $	–	1542.94	12,317.82	12,733.20	12,836.90
% saving	–	4.08	32.58	33.69	33.96
Minimum voltage, p.u.	0.9091 (bus 65)	0.9504 (bus 61)	0.9287 (bus 65)	0.9298 (bus 65)	0.9313 (bus 65)
Maximum voltage, p.u.	0.9999 (bus 2)	0.9999 (bus 2)	0.9999 (bus 2)	0.9999 (bus 2)	0.9999 (bus 2)

a maximum of 0.9999 2nd. Table 4.14 shows a detailed comparison between the results obtained by the proposed method and the results of the DE, ABC and GSA algorithms.

4.6.2.4 Statistical Analysis

In order to validate the results a statistical analysis between the different methods is performed and the results are illustrated in Table 4.15.

Table 4.15 Statistical analysis

Wilcoxon	DE-ICSA	ABC-ICSA	GSA-ICSA
10-Bus	2.5576×10^{-34}	2.4788×10^{-34}	2.5566×10^{-34}
33-Bus	6.1019×10^{-34}	3.4570×10^{-32}	7.6490×10^{-24}
69-Bus	1.0853×10^{-29}	8.6857×10^{-28}	3.6802×10^{-20}

4.7   Conclusions

In this chapter, an improved version of the CSA method is presented to solve complex high multi-modal optimization problems of energy: Identification of induction motors and capacitor allocation in distribution networks. In the new algorithm, two features of the original CSA are modified: (I) the awareness probability (AP) and (II) the random perturbation. With the purpose to enhance the exploration–exploitation ratio the fixed awareness probability (AP) value is replaced (I) by a dynamic awareness probability (DAP), which is adjusted according to the fitness value of each candidate solution. The Lévy flight movement is also incorporated to enhance the search capacities of the original random perturbation (II) of CSA. With such adaptations, the new approach preserves solution diversity and improves the convergence to difficult high multi-modal optima.

In order to evaluate its performance, the proposed algorithm has been compared with other popular search algorithms such as the DE, ABC and GSA. The results demonstrate the high performance of the proposed method in terms of accuracy and robustness.

Appendix A: Systems Data

See Tables 4.16, 4.17 and 4.18.

Table 4.16 10-bus test system data

Line No.	From bus i	To bus $i + 1$	R (Ω)	X (Ω)	P_L (kW)	Q_L (kVAr)
1	1	2	0.1233	0.4127	1840	460
2	2	3	0.0140	0.6051	980	340
3	3	4	0.7463	1.2050	1790	446
4	4	5	0.6984	0.6084	1598	1840
5	5	6	1.9831	1.7276	1610	600
6	6	7	0.9053	0.7886	780	110
7	7	8	2.0552	1.1640	1150	60
8	8	9	4.7953	2.7160	980	130
9	9	10	5.3434	3.0264	1640	200

Table 4.17 33-bus test system data

Line No.	From bus i	To bus $i + 1$	R (Ω)	X (Ω)	P_L (kW)	Q_L (kVAr)
1	1	2	0.0922	0.0477	100	60
2	2	3	0.4930	0.2511	90	40
3	3	4	0.3660	0.1864	120	80
4	4	5	0.3811	0.1941	60	30
5	5	6	0.8190	0.7070	60	20
6	6	7	0.1872	0.6188	200	100
7	7	8	1.7114	1.2351	200	100
8	8	9	1.0300	0.7400	60	20
9	9	10	1.0400	0.7400	60	20
10	10	11	0.1966	0.0650	45	30
11	11	12	0.3744	0.1238	60	35
12	12	13	1.4680	1.1550	60	35
13	13	14	0.5416	0.7129	120	80
14	14	15	0.5910	0.5260	60	10
15	15	16	0.7463	0.5450	60	20
16	16	17	1.2890	1.7210	60	20
17	17	18	0.7320	0.5740	90	40
18	2	19	0.1640	0.1565	90	40
19	19	20	1.5042	1.3554	90	40
20	20	21	0.4095	0.4784	90	40
21	21	22	0.7089	0.9373	90	40
22	3	23	0.4512	0.3083	90	50
23	23	24	0.8980	0.7091	420	200
24	24	25	0.8960	0.7011	420	200
25	6	26	0.2030	0.1034	60	25
26	26	27	0.2842	0.1447	60	25
27	27	28	1.0590	0.9337	60	20
28	28	29	0.8042	0.7006	120	70
29	29	30	0.5075	0.2585	200	600
30	30	31	0.9744	0.9630	150	70
31	31	32	0.3105	0.3619	210	100
32	32	33	0.3410	0.5302	60	40

Table 4.18 69-bus test system data

Line No.	From bus i	To bus $i + 1$	R (Ω)	X (Ω)	P_L (kW)	Q_L (kVAr)
1	1	2	0.00050	0.0012	0.00	0.00
2	2	3	0.00050	0.0012	0.00	0.00
3	3	4	0.00150	0.0036	0.00	0.00
4	4	5	0.02510	0.0294	0.00	0.00
5	5	6	0.36600	0.1864	2.60	2.20
6	6	7	0.38100	0.1941	40.40	30.00
7	7	8	0.09220	0.0470	75.00	54.00
8	8	9	0.04930	0.0251	30.00	22.00
9	9	10	0.81900	0.2707	28.00	19.00
10	10	11	0.18720	0.0619	145.00	104.00
11	11	12	0.71140	0.2351	145.00	104.00
12	12	13	1.03000	0.3400	8.00	5.00
13	13	14	1.04400	0.3400	8.00	5.00
14	14	15	1.05800	0.3496	0.00	0.00
15	15	16	0.19660	0.0650	45.00	30.00
16	16	17	0.37440	0.1238	60.00	35.00
17	17	18	0.00470	0.0016	60.00	35.00
18	18	19	0.32760	0.1083	0.00	0.00
19	19	20	0.21060	0.0690	1.00	0.60
20	20	21	0.34160	0.1129	114.00	81.00
21	21	22	0.01400	0.0046	5.00	3.50
22	22	23	0.15910	0.0526	0.00	0.00
23	23	24	0.34630	0.1145	28.00	20.00
24	24	25	0.74880	0.2475	0.00	0.00
25	25	26	0.30890	0.1021	14.00	10.00
26	26	27	0.17320	0.0572	14.00	10.00
27	3	28	0.00440	0.0108	26.00	18.60
28	28	29	0.06400	0.1565	26.00	18.60
29	29	30	0.39780	0.1315	0.00	0.00
30	30	31	0.07020	0.0232	0.00	0.00
31	31	32	0.35100	0.1160	0.00	0.00
32	32	33	0.83900	0.2816	14.00	10.00
33	33	34	1.70800	0.5646	19.50	14.00
34	34	35	1.47400	0.4873	6.00	4.00
35	3	36	0.00440	0.0108	26.00	18.55
36	36	37	0.06400	0.1565	26.00	18.55

(continued)

Table 4.18 (continued)

Line No.	From bus i	To bus $i + 1$	R (Ω)	X (Ω)	P_L (kW)	Q_L (kVAr)
37	37	38	0.10530	0.1230	0.00	0.00
38	38	39	0.03040	0.0355	24.00	17.00
39	39	40	0.00180	0.0021	24.00	17.00
40	40	41	0.72830	0.8509	1.20	1.00
41	41	42	0.31000	0.3623	0.00	0.00
42	42	43	0.04100	0.0478	6.00	4.30
43	43	44	0.00920	0.0116	0.00	0.00
44	44	45	0.10890	0.1373	39.22	26.30
45	45	46	0.00090	0.0012	39.22	26.30
46	4	47	0.00340	0.0084	0.00	0.00
47	47	48	0.08510	0.2083	79.00	56.40
48	48	49	0.28980	0.7091	384.70	274.50

References

1. Prakash, V., Baskar, S., Sivakumar, S., Krishna, K.S.: A novel efficiency improvement measure in three-phase induction motors, its conservation potential and economic analysis. Energy. Sustain. Dev. **12**, 78–87 (2008). https://doi.org/10.1016/S0973-0826(08)60431-7
2. Saidur, R.: A review on electrical motors energy use and energy savings. Renew. Sustain. Energy Rev. **14**, 877–898 (2010). https://doi.org/10.1016/j.rser.2009.10.018
3. Perez, I., Gomez-Gonzalez, M., Jurado, F.: Estimation of induction motor parameters using shuffled frog-leaping algorithm. Electr. Eng. **95**, 267–275 (2013). https://doi.org/10.1007/s00202-012-0261-7
4. Sakthivel, V.P., Bhuvaneswari, R., Subramanian, S.: Artificial immune system for parameter estimation of induction motor. Expert Syst. Appl. **37**, 6109–6115 (2010). https://doi.org/10.1016/j.eswa.2010.02.034
5. Lee, S., Grainger, J.: Optimum placement of fixed and switched capacitors on primary distribution feeders. IEEE Trans. Power Appar. Syst. 345–352 (1981). https://doi.org/10.1109/tpas.1981.316862
6. Zidar, M., Georgilakis, P.S., Hatziargyriou, N.D., Capuder, T., Škrlec, D.: Review of energy storage allocation in power distribution networks: applications, methods and future research. IET Gener. Trans. Distrib. **3**, 645–652 (2016). https://doi.org/10.1049/iet-gtd.2015.0447
7. Grainger, J., Lee, S.: Optimum size and location of shunt capacitors for reduction of losses on distribution feeders. IEEE Trans. Power Appar. Syst. **3**, 1105–1118 (1981). https://doi.org/10.1109/TPAS.1981.316577
8. Neagle, N.M., Samson, D.R.: Loss reduction from capacitors installed on primary feeders. Trans. Am. Inst. Electr. Eng. Part III Power Appar. Syst **75**, 950–959 (1956). https://doi.org/10.1109/AIEEPAS.1956.4499390
9. Dura, H.: Optimum number, location, and size of shunt capacitors in radial distribution feeders a dynamic programming approach. IEEE Trans. Power Appar. Syst. **9**, 1769–1774 (1968). https://doi.org/10.1109/TPAS.1968.291982
10. Hogan, P.M., Rettkowski, J.D., Bala, J.L.: Optimal capacitor placement using branch and bound. In: Proceedings of the 37th Annual North American Power Symposium, Ames, IA, USA, 25 October 2005; pp. 84–89. https://doi.org/10.1109/naps.2005.1560506

11. Mekhamer, S.F., El-Hawary, M.E., Soliman, S.A., Moustafa, M.A., Mansour, M.M.: New heuristic strategies for reactive power compensation of radial distribution feeders. IEEE Trans. Power Deliv. **17**, 1128–1135 (2002). https://doi.org/10.1109/TPWRD.2002.804004
12. Da Silva, I.C., Carneiro, S., de Oliveira, E.J., de Souza Costa, J., Pereira, J.L.R., Garcia, P.A.N.: A heuristic constructive algorithm for capacitor placement on distribution systems. IEEE Trans. Power Syst. **23**, 1619–1626 (2008). https://doi.org/10.1109/TPWRS.2008.2004742
13. Chis, M., Salama, M.M.A., Jayaram, S.: Capacitor placement in distribution systems using heuristic search strategies. IEE Proc. Gener. Transm. Distrib. **144**, 225 (1997). https://doi.org/10.1049/ip-gtd:19970945
14. Askarzadeh, A.: Capacitor placement in distribution systems for power loss reduction and voltage improvement: a new methodology. IET Gener. Transm. Distrib. **10**, 3631–3638 (2016). https://doi.org/10.1049/iet-gtd.2016.0419
15. Ng, H.N., Salama, M.M.A., Chikhani, A.Y.: Classification of capacitor allocation techniques. IEEE Trans. Power Deliv. **15**, 387–392 (2000). https://doi.org/10.1109/61.847278
16. Jordehi, A.R.: Optimisation of electric distribution systems: a review. Renew. Sustain. Energy Rev. **51**, 1088–1100 (2015). https://doi.org/10.1016/j.rser.2015.07.004
17. Aman, M.M., Jasmon, G.B., Bakar, A.H.A., Mokhlis, H., Karimi, M.: Optimum shunt capacitor placement in distribution system—a review and comparative study. Renew. Sustain. Energy Rev. **30**, 429–439 (2014). https://doi.org/10.1016/j.rser.2013.10.002
18. Avalos, O., Cuevas, E., Gálvez, J.: Induction motor parameter identification using a gravitational search algorithm. Computers **5**, 6 (2016). https://doi.org/10.3390/computers5020006
19. Mohamed Shuaib, Y., Surya Kalavathi, M., Christober Asir Rajan, C.: Optimal capacitor placement in radial distribution system using gravitational search algorithm. Int. J. Electr. Power Energy Syst. **64**, 384–397 (2015). https://doi.org/10.1016/j.ijepes.2014.07.041
20. Sakthivel, V.P., Bhuvaneswari, R., Subramanian, S.: An accurate and economical approach for induction motor field efficiency estimation using bacterial foraging algorithm. Measurement **44**, 674–684 (2011). https://doi.org/10.1016/j.measurement.2010.12.008
21. Devabalaji, K.R., Ravi, K., Kothari, D.P.: Optimal location and sizing of capacitor placement in radial distribution system using Bacterial Foraging Optimization Algorithm. Int. J. Electr. Power Energy Syst. **71**, 383–390 (2015). https://doi.org/10.1016/j.ijepes.2015.03.008
22. Picardi, C., Rogano, N.: Parameter identification of induction motor based on particle swarm optimization. In: Proceedings of the International Symposium on Power Electronics, Electrical Drives, Automation and Motion, Taormina, Italy, 23–26 May 2006; pp. 968–973. https://doi.org/10.1109/speedam.2006.1649908
23. Singh, S.P., Rao, A.R.: Optimal allocation of capacitors in distribution systems using particle swarm optimization. Int. J. Electr. Power Energy Syst. **43**, 1267–1275 (2012). https://doi.org/10.1016/j.ijepes.2012.06.059
24. Prakash, K., Sydulu, M.: Particle swarm optimization based capacitor placement on radial distribution systems. In: Proceedings of the 2007 IEEE Power Engineering Society General Meeting, Tampa, FL, USA, 24–28 June 2017; pp. 1–5. https://doi.org/10.1109/pes.2007.386149
25. Lee, C.-S., Ayala, H.V.H., dos Santos Coelho, L.: Capacitor placement of distribution systems using particle swarm optimization approaches. Int. J. Electr. Power Energy Syst. 64, 839–851 (2015). https://doi.org/10.1016/j.ijepes.2014.07.069
26. Alonge, F., D'Ippolito, F., Ferrante, G., Raimondi, F.: Parameter identification of induction motor model using genetic algorithms. IEE Proc. Control Theory Appl. **145**, 587–593 (1998)
27. Swarup, K.: Genetic algorithm for optimal capacitor allocation in radial distribution systems. In: Proceedings of the 6th WSEAS international conference on evolutionary, Lisbon, Portugal, 16–18 June 2005
28. Ursem, R., Vadstrup, P.: Parameter identification of induction motors using differential evolution. In: Proceedings of the 2003 Congress on Evolutionary Computation, Canberra, ACT, Australia, 8–12 December 2003
29. Su, C., Lee, C.: Modified differential evolution method for capacitor placement of distribution systems. In: Proceedings of the 2002 Asia Pacific. IEEE Transmission and Distribution Conference and Exhibition, Yokohama, Japan, 6–10 October 2002

30. Chiou, J., Chang, C., Su, C.: Capacitor placement in large-scale distribution systems using variable scaling hybrid differential evolution. Int. J. Electr. Power **28**, 739–745 (2006)
31. Huang, Y.-C., Yang, H.T., Huang, C.L.: Solving the capacitor placement problem in a radial distribution system using Tabu Search approach. IEEE Trans. Power Syst. **11**, 1868–1873 (1996). https://doi.org/10.1109/59.544656
32. Olaimaei J, Moradi M, Kaboodi T.: A new adaptive modified firefly algorithm to solve optimal capacitor placement problem. Electr Power Distribution Networks Conference (2013). https://doi.org/10.1109/edp.2013.6565962
33. Askarzadeh, A.: A novel metaheuristic method for solving constrained engineering optimization problems: crow search algorithm. Comput. Struct. **169**, 1–12 (2016). https://doi.org/10.1016/j.compstruc.2016.03.001
34. Oliva, D., Hinojosa, S., Cuevas, E., Pajares, G., Avalos, O., Gálvez, J.: Cross entropy based thresholding for magnetic resonance brain images using Crow Search Algorithm. Expert Syst. Appl. **79**, 164–180 (2017). https://doi.org/10.1016/j.eswa.2017.02.042
35. Liu, D., Liu, C., Fu, Q., Li, T., Imran, K., Cui, S., Abrar, F.: ELM evaluation model of regional groundwater quality based on the crow search algorithm. Ecol. Indic. **81**, 302–314 (2017)
36. Emery, N.J., Clayton, N.S.: The mentality of crows: convergent evolution of intelligence in corvids and apes. Science **306**, 1903–1907 (2004)
37. Holzhaider, J.C., Hunt, G.R., Gray, R.D.: Social learning in New Caledonian crows. Learn. Behav. **38**, 206–219 (2010). https://doi.org/10.3758/LB.38.3.206
38. Rajput, S., Parashar, M., Dubey, H.M., Pandit, M.: Optimization of benchmark functions and practical problems using Crow Search Algorithm. In: Proceedings of the 2016 Fifth International Conference on Eco-friendly Computing and Communication Systems (ICECCS), Bhopal, India, 8–9 December 2016; pp. 73–78. https://doi.org/10.1109/eco-friendly.2016.789 3245
39. Karaboga, D., Basturk, B.: A powerful and efficient algorithm for numerical function optimization: artificial bee colony (ABC) algorithm. J. Glob. Optim. **39**, 459–471 (2007)
40. Baronchelli, A., Radicchi, F.: Lévy flights in human behavior and cognition. Chaos, Solitons Fractals **56**, 101–105 (2013)
41. Sotolongo-Costa, O., Antoranz, J.C., Posadas, A., Vidal, F., Vazquez, A.: Levy Flights and Earthquakes. Geophys. Res. Lette. **27**, 1965–1968 (2000)
42. Shlesinger, M.F., Zaslavsky, G.M., Frisch, U.: Lévy Flights and Related Topics in Physics. Springer, Berlin, Germany (1995)
43. Yang, X.-S., Ting, T.O., Karamanoglu, M.: Random Walks, Lévy Flights, Markov Chains and Metaheuristic Optimization. Springer: Dordrecht, The Netherlands, 2013; pp. 1055–1064. https://doi.org/10.1007/978-94-007-6516-0_116
44. Yang, X.-S.: Nature-Inspired Metaheuristic Algorithms. Luniver Press, Beckington, UK (2010)
45. Venkatesh, B., Ranjan, R.: Data structure for radial distribution system load flow analysis. IEE Proc. Gener. Transm. Distrib. **150**, 101–106 (2003)
46. Chen, T., Chen, M., Hwang, K.: Distribution system power flow analysis-a rigid approach. IEEE Trans. Power Deliv. **6**, 1146–1152 (1991)
47. Ghosh, S., Das, D.: Method for load-flow solution of radial distribution networks. Gener. Trans. Distrib. **146**, 641–648 (1999)
48. Rao, R.S., Narasimham, S.V.L., Ramalingaraju, M.: Optimal capacitor placement in a radial distribution system using Plant Growth Simulation Algorithm. Int. J. Electr. Power Energy Syst. **33**, 1133–1139 (2011). https://doi.org/10.1016/j.ijepes.2010.11.021
49. Jamadi, M., Merrikh-Bayat, F.: New Method for Accurate Parameter Estimation of Induction Motors Based on Artificial Bee Colony Algorithm. arXiv (2014). arXiv:1402.4423
50. Grainger, J.J., Lee, S.H.: Capacity release by shunt capacitor placement on distribution feeders: a new voltage-dependent model. IEEE Trans. Power Appar. Syst. **5**, 1236–1244 (1982). https://doi.org/10.1109/TPAS.1982.317385
51. Baran, M.E., Wu, F.F.: Network reconfiguration in distribution systems for loss reduction and load balancing. IEEE Trans. Power Deliv. **4**, 1401–1407 (1989). https://doi.org/10.1109/61.25627

52. Baran, M., Wu, F.F.: Optimal sizing of capacitors placed on a radial distribution system. IEEE Trans. Power Deliv. **4**, 735–743 (1989). https://doi.org/10.1109/61.19266
53. Cuevas, E., Block-matching algorithm based on harmony search optimization for motion estimation, Appl. Intell. **39**(1), 165–183 (2013)
54. Díaz-Cortés, M.-A., Ortega-Sánchez, N., Hinojosa, S., Cuevas, E., Rojas, R., Demin, A.: A multi-level thresholding method for breast thermograms analysis using Dragonfly algorithm. Infrared Phys. Technol. **93**, 346–361 (2018)
55. Díaz, P., Pérez-Cisneros, M., Cuevas, E., Hinojosa, S., Zaldivar, D.: An im-proved crow search algorithm applied to energy problems. Energies **11**(3), 571 (2018)

Chapter 5
ANFIS-Hammerstein Model
for Nonlinear Systems Identification
Using GSA

Abstract The nature of many real problems in the world is nonlinear type, and identifying their plants and processes symbolizes a challenging task. Nowadays, the block-structure systems, such as the Hammerstein model, are among the most current nonlinear systems. The main characteristic of a Hammerstein model is that its architecture is made up of two blocks; a linear dynamic model preceded by a static nonlinear. The adaptive neuro-fuzzy inference system (ANFIS) is a robust scheme that incorporates a two parts structural design; a nonlinear rule-based and a linear system. In this chapter, it is proposed a new scheme based on the Hammerstein block-structure model for nonlinear system identification. The methodology introduced takes benefit of the correspondence between the ANFIS and Hammerstein structure to couple them and model nonlinear systems. The Gravitational Search Algorithm (GSA) is incorporated to the methodology to identify the model system parameters. The GSA, compared to similar optimization algorithms, achieves more reliable performance in multimodal problems, avoiding being trapped in premature solutions that are not optimal. To test and validate the effectiveness of the methodology, it has been tested over several models and compared with related works in literature showing a higher accuracy in the results.

5.1 Introduction

The behavior of most useful systems is nonlinear in nature. The identification problem of nonlinear systems is recognized as a hard and complex task, for this reason the research community has been interested in the development of new system modeling methodologies. Hammerstein models [1] is a relevant scheme in the modeling area, this popularity has been reached by their efficiency and accuracy to model complex nonlinear systems. Hammerstein is a block-structure model, which its most representative feature is the interconnection of a nonlinear block function followed by a dynamic linear block. The simplicity in structure allows an ease understanding of Hammerstein system which has led it to promote its use in a diverse nonlinear

© Springer Nature Switzerland AG 2021
E. Cuevas et al., *Metaheuristic Computation: A Performance Perspective*,
Intelligent Systems Reference Library 195,
https://doi.org/10.1007/978-3-030-58100-8_5

modeling application such as control systems [5], solid oxide fuel cells [2], power amplifier pre-distortion [4], PH process [6], ultrasonic motors [3], and so on.

The process to model a system through a Hammerstein approach is conducted considering the design of both linear and nonlinear blocks, the nonlinear being the critical part of the procedure. Different kinds of adaptive networks can approximate the stationary, nonlinear block. In literature, find schemes such as the multi-layer perceptron. (MLP) [7], a Single hidden layer feed-forward neural network (SFLNs) [8] or a Functional link artificial neural network (FLANN) [9, 10]. The linear block is commonly described by an adaptive in infinite impulse response (IIR) filter.

The adaptive neuro-fuzzy inference system (ANFIS) [11] is a relevant type of adaptable network for nonlinear systems that have reached notoriety as a soft computing method. ANFIS mixes the strengths of two areas of artificial intelligence, on the one hand, the learning ability of adaptive networks through the partial truth reasoning of the Fuzzy system Takagi-Sugeno [12] and the other the inference using Fuzzy approach to obtain an accurate and generalization relationship of input and output pairs. Based on that, the ANFIS architecture can be sectioned into two parts, the adaptive network for the nonlinear part and the Fuzzy inference system for the linear block. Due to the flexibility in its architecture, ANFIS has been applied to a wide variety of problems involving time-series prediction [13], estimation of wind speed profile [14], fault detection of an industrial steam turbine [15], representation of a nonlinear liquid-level system [16], modeling of solid oxide fuel cell [17], and so on.

An essential step in identifying a Hammerstein model is the estimation of the weights and parameters of the adaptive nonlinear system and the linear system, respectively. This procedure becomes an optimization problem that groups the weights and parameters of the nonlinear and linear system as a multidimensional decision variable. With such consideration, the search space of the optimization problem tends to be multimodal, increasing the complexity of the objective function and its solution. Generally, traditional deterministic techniques based on gradient descent or the least mean square are considered by most of the approaches to perform the estimation of Hammerstein parameters.

Due to certain limitations of deterministic techniques, other approaches are integrated to solve the Hammerstein model identification problem. One such approach is evolutionary computing techniques that have been shown to produce models with higher confidence in terms of exactitude and robustness than deterministic methodologies [18]. Two examples of the methodologies used in model identification are Particle Swarm Optimization (PSO) [19] and Differential Evolution (DE) [20]. Although these methods perform adequate model identification results, they have a poor balance between diversification-intensification, which generates suboptimal results.

On the other hand, a potent approach of the evolutionary computing framework is the the Gravitational Search Algorithm (GSA) [21]. The general concept of GSA is based on the laws of gravity and mass interactions. The GSA, in comparison with the most of used evolutionary algorithms in the literature, GSA has a better profile in complex problems with multimodal solutions, having the ability of being able to go out of optimal local [22, 23]. These attributes have led it to be applied to solve some engineering areas such energy [24], image processing [25] and machine learning [26].

In this chapter, an ANFIS-Hammerstein model for nonlinear systems identification is introduced. In the proposed methodology, the ANFIS scheme is adapted to model a Hammerstein system. Under such adaptation, there is a direct correspondence between the Hammerstein blocks to the ANFIS architecture. Considering the benefit of the connection between the both schemes, the proposed work combines the advanced modeling capabilities of ANFIS and the prominent multimodal features of the nature-inspired GSA to estimate the parameters values of a Hammerstein model, through the ANFIS structure. To test and validate the effectiveness of the methodology, it has been tested over several models and compared with related works in literature showing a higher accuracy in the results.

The chapter is organized as follows. In Sect. 5.2, the ANFIS structure and the GSA algorithm are described. Section 5.3 formulates the proposed identification scheme. Section 5.4 shows the experimental results. Finally, some conclusions are discussed in Sect. 5.5.

5.2 Background

5.2.1 Hybrid ANFIS Models

In ANFIS is a fusion of two fields of computational intelligence developed through the combination of the learning capacity of neural networks and the inferential capacity of fuzzy logic. The attractive hybridization scheme of ANFIS has led it to be used to identify and model a wide range of systems [13–17]. In the learning stage, the estimation of ANFIS parameters is carried out by employing an optimization algorithm. Commonly, the gradient-based and evolutionary are the two methods that are employed for this task [27].

Even though both methods are employed in the determination of the ANFIS parameters, the evolutionary algorithms are preferred over the gradient-based mainly because these last ones regularly find sub-optimal results. In contrast, evolutionary methods are characterized by achieving better results at global optimum in solving complex optimization problems [27]. One factor why these evolutionary algorithms are successful in finding optimal solutions is their ability to simultaneously exploration-exploitation over the search space.

Nowadays, several methods reported in the literature include the combined use of ANFIS and evolutionary approaches to identify and model systems. Some of them are mentioned below.

In [28], Sarkheyli et al. estimated the ANFIS parameters via a variation of the Genetic Algorithm (GA). Wei, in [29], combined a composite model based on GA and ANFIS to determine the stock value in Taiwan. Rini et al. [30] examined the capabilities of the response and accuracy of ANFIS modeling when it is training by Particle Swarm Optimization (PSO) algorithm. Wang and Ning, in [31], joined the PSO and ANFIS to predicted time series behavior. Liu et al. [32] applied an enhanced variant of Quantum-behaved Particle Swarm Optimization (QPSO) to calc the ANFIS model parameters. In [33], Catalão et al. generated a combination model integrating PSO, wavelet transform, and ANFIS for short-term electricity cost forecast. Suja Priyadharsini et al. [34] proposed a method for the removal of artifacts in Electroencephalography (EEG) signals, through ANFIS and Artificial Immune System (AIS). In [35], Gunasekaran and Ramaswami fused the ANFIS system and an AIS algorithm to forecast the projected value of the National Stock Exchange (NSE) of India. Asadollahi-Baboli [36] used the ANFIS in association with a variation of ANt Colony Optimization (ACO) as a predictor of chemical particles. In [37], Teimouri and Baseri introduced an ANFIS model to estimate the parameter relationship in an electrical discharge machining (EDM) method. In the proposal, the Continuous Ant Colony Optimization (CACO) method is applied as a training algorithm for the ANFIS structure. Varshini et al. [38] conducted the ANFIS with Bacterial Foraging Optimization Algorithm (BFOA) as a learning method to estimate ANFIS parameters. In [39], Bhavani et al. suggested an ANFIS-BFOA approach to control the load frequency in the competitive electricity business. Karaboga and Kaya [27] proposed a variant of Artificial Bee Colony (ABC) algorithm for training the ANFIS scheme. In [40], Azarbad et al. used the methodology of ANFIS to model Global Positioning System (GPS) information. The ABC in combinations with Radial Basis Functions (RBF) are the involved for the learning task. Rani and Sankaragomathi [41] integrated ANFIS with the metaheuristic optimization algorithm cuckoo search (CS) to design a PID controller. In [42], Chen and Do modeled student educational achievement, through using ANFIS and CS as learning algorithms. Khazraee et al. [43] modeled a reactive batch distillation process using used ANFIS and the Differential Evolution (DE) algorithm. In [44], Zangeneh et al. integrated two methods for the training process in the identification of nonlinear models via ANFIS. The parameters of the nonlinear section are approximated by DE, while the linear block is estimated by a classic gradient-based method. Afifi et al. [45] exposed a method to expects the cellular phone malware existence. the method includes a merge of ANFIS scheme and PSO algorithm.

The overall methods conceive the ANFIS approach as a black-box representation, while the computational techniques are employed as the core of the learning process to determine its parameters. In contrast to referred approaches, in the proposed methodology, the ANFIS structure is employed directly as a Hammerstein architecture to model nonlinear plants. Due to their structure relation, the plant parameters of the Hammerstein system can be efficiently adapted to the ANFIS design. Hence, the process of adapting an ANFIS model matches with the identification of the Hammerstein structure. Following this perspective, the result is a gray system where the parameters of the ANFIS model assemble with those of the Hammerstein plant.

On the other hand, the application of computational intelligent techniques has previously been applied to identify the Hammerstein model. Some works reported in the literature that includes some evolutionary method to optimize the estimation process of Hammerstein model are those presented by Duwa [19], Tang et al. [8], Mete et al. [20], Cui et al. [7], and Gotmare et al. [10]. These works achieve interesting results combining an evolutionary method with a soft computing component, however, they still address the problem as a black-box model limiting the opportunity to perceive the inside parameters directly. Additionally, the evolutionary methods used in these works commonly have certain limitations to obtain the global optimum due to the multimodal problem surface.

5.2.2 Adaptive Neuro-Fuzzy Inference System (ANFIS)

ANFIS [11] is a neuro-fuzzy system that synergizes the adaptability of neural networks and the Takagi-Sugeno fuzzy reasoning system. According to this architecture, ANFIS can be divided in a nonlinear block and a linear system. The nonlinear block involves the antecedent of the fuzzy inference system, while the linear system represents the linear functions of the consequent (see Fig. 5.1). Structurally, the ANFIS model involves five layers. Each one of these layers has different purposes, which are described below.

Layer 1: Every node i in this layer specifies the degree to which an input u satisfies the linguistic concept characterized by A_i.

$$O_i^1 = \mu_{A_i}(u) \tag{5.1}$$

This correspondence is calculated by a membership function modeled as follows:

$$\mu_{A_i}(u) = e^{-\frac{(u-o_i)}{(2s_i)^2}} \tag{5.2}$$

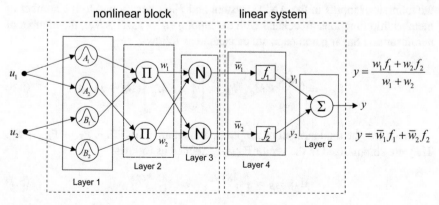

Fig. 5.1 ANFIS structure

where $\{\sigma_i, s_i\}$ represents the adaptable parameters of the membership function. This parameter set is known as the premise parameters.

Layer 2: Here, the final output O_i^2 $\left(O_i^2 = w_i\right)$ of each node i (Π) represents the firing strength of its corresponding fuzzy rule.

Layer 3: Each node i (N) in this layer calculates the normalized firing strength formulated as follows:

$$O_i^3 = \bar{w} = \frac{w_i}{\sum_{j=1}^{h} w_j} \tag{5.3}$$

where h represents the number of nodes in this layer.

Layer 4: Every node i in this layer corresponds to an adaptive processing unit that implements a linear function defined as follows:

$$O_i^4 = \bar{w}_i f_i = \bar{w}_i (p_i u_i + q_i u_2 + r\_i) \tag{5.4}$$

where $\{p_i, q_i, r_i\}$ represent the adjustable parameters that determine the function response. The parameter set of this layer is referred to as the consequence elements. In general, f_i is a function which can be modeled by any linear function as long as it can appropriately describe the behavior of the system within the fuzzy region specified by the antecedent of rule i.

Layer 5: This node (Σ) calculates the overall output y as the sum of all incoming signals:

$$y = O_i^5 = \sum_i \bar{w}_i f_i = \frac{\sum_i w_i f_i}{\sum_i w_i} \tag{5.5}$$

ANFIS divides its internal parameters P_T into nonlinear P_{NL} and linear P_L. The nonlinear parameters P_{NL} represent the elements $\{\sigma_i, s_i\}$ while the elements $\{p_i, q_i, r_i\}$ correspond to the linear parameters P_L. Assuming that N_u represents the number of inputs in the ANFIS system and FM_q correspond to the number of membership functions associated with the input q ($q = 1, \ldots, N_u$), the number of nonlinear and linear parameters are calculated as follows:

$$P_{NL} = 2 \sum_{q=1}^{N_u} FM_q, \quad P_L = \left[\prod_{q=1}^{N_u} FM_q \right] (N_u + 1) \tag{5.6}$$

Therefore, the total number of adaptive parameters in ANFIS is $P_T = P_{NL} + P_L$. They are summarized in the vector E_{ANFIS} defined as follows:

$$\mathrm{E}_{ANFIS} = \left\{ e_1^{ANFIS}, \ldots, e_{P_T}^{ANFIS} \right\} \tag{5.7}$$

ANFIS uses a hybrid learning algorithm that combines the steepest descent technique for P_{NL} and the least-squares method for P_L. This hybrid learning process presents acceptable results; however, it frequently obtains sub-optimal solutions as a consequence of its inability to face multimodal functions.

5.2.3 Gravitational Search Algorithm (GSA)

The Gravitational search algorithm is an evolutionary technique based on the laws of gravity, proposed in 2009 by Rashedi [21]. GSA is a relatively new optimization technique that has been gaining popularity due to its good performance in complex tasks, avoiding effectively sub-optimal solutions in multimodal problems. In GSA, each solution emulates a mass that has a gravitational force, which allows it to attract other masses. In the GSA context, the quality of each solution is evaluated by an objective function. GSA is a population approach that considers N different candidate solutions or masses, where each mass represents a d-dimensional vector formulated as follows:

$$m_i(t) = \{m_i^1, \ldots, m_i^d\}; \quad i = 1, \ldots, N \tag{5.8}$$

In GSA, at time t, the attraction force acting between mass i and mass j in the h variable ($h \in 1, \ldots, d$) is defined as follows:

$$F_{ij}^h(t) = G(t) \frac{M_i(t)M_j(t)}{R_{ij}(t) + \varepsilon} \left(m_j^h(t) - m_i^h(t)\right) \tag{5.9}$$

where $G(t)$ represents the gravitational constant at time t, M_i and M_j are the gravitational mass of solutions i and j, respectively. R_{ij} symbolizes the Euclidian distance between the i-th and j-th solutions, and ε is a small constant. To maintain the equilibrium between the exploration and exploitation, GSA alters the attraction force $G(t)$ between solutions during the evolution process. Under such conditions, the final force acting over the solution i is described as follows:

$$F_i^h(t) = \sum_{j=1, j \neq i}^{N_u} rand * F_{ij}^h(t) \tag{5.10}$$

where $rand$ symbolizes a random number in the interval [0, 1]. The acceleration of the candidate solution i is given as follows:

$$a_i^h(t) = \frac{F_i^h(t)}{M_i(t)} \tag{5.11}$$

Therefore, the new position and velocity of each candidate solution i can be computed as follows:

$$v_i^h(t+1) = rand * v_i^h(t) + a_i^h(t)$$
$$m_i^h(t+1) = m_i^h(t) + v_i^h(t) \tag{5.12}$$

At each iteration, the Gravitational masses are calculated by using the cost function which determines the quality of the candidate solutions. Therefore, the gravitational masses are updated by using the following equations:

$$p_i(t)_i^h = \frac{f(m_i(t)) - worst(t)}{best(t) - worst(t)} \tag{5.13}$$

$$M_i(t) = \frac{p_i(t)}{\sum_{l=1}^{N} p_l(t)} \tag{5.14}$$

where $f(.)$ corresponds to the objective function, and $worst(t)$ and $best(t)$ represent the worst and best fitness values at time (t) respectively.

5.3 Hammerstein Model Identification by Using GSA

A Hammerstein approach essentially combines a nonlinear static operator **NS** with a dynamic linear system **LS** in order to model a nonlinear plant **P**. Under this approach, the nonlinearity is modeled by the static function **NS** while the linear system **LS** imposes the dynamic behavior.

The nonlinear static system **NS** consists of a nonlinear function that introduces the nonlinearity in the Hammerstein model. **NS** receives an input $u(k)$ and delivers an output $x(k)$ which is transformed by a nonlinear operator $\beta(\cdot)$ such that $x(k) = \beta(u(k))$. Under these conditions, the input $u(k)$ of **NS** represents the input of the Hammerstein model while the output $x(k)$ of **NS** is the input of the linear system **LS** of the Hammerstein structure. Figure 5.2 shows a graphical representation of the Hammerstein model.

Fig. 5.2 Graphical representation of the Hammerstein model

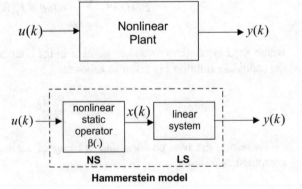

The input-output modeling of the linear part **LS** in the Hammerstein structure is typically characterized as an adjustable infinite impulse response (IIR) filter defined as follows:

$$y(k) = \frac{B(z)}{A(z)} x(k-1) + \frac{C(z)}{A(z)} \delta(k)$$
$$A(z^{-1}) = 1 + a_1 z^{-1} + \cdots + a_P z^{-P}$$
$$B(z^{-1}) = b_0 + b_1 z^{-1} + \cdots + b_L z^{-L-1}$$
$$C(z^{-1}) = c_0 + c_1 z^{-1} + \cdots + c_S z^{-S-1} \tag{5.15}$$

where P, L and S represent the polynomial order of $A(z)$, $B(z)$ and $C(z)$, respectively. In the linear system modeled by Eq. (5.15), $x(k)$ and $y(k)$ symbolize the input and output of the plant while $\delta(k)$ is the noise signal that represents the unmodeled effects.

In this paper, a nonlinear system identification method, based on the Hammerstein model, is introduced. In the proposed scheme, the system is modeled through the adaptation of an ANFIS scheme, taking advantage of the similarity between it and the Hammerstein model. The proposed model is shown in Fig. 5.3. The structure of ANFIS consists of an adaptable Takagi-Sugeno fuzzy system. According to this architecture, ANFIS can be divided into a nonlinear block and a linear system. The nonlinear block involves the antecedent of the fuzzy inference system, while the linear system represents the linear functions of the consequent. The output of the nonlinear part $x(k)$ represents the input of the linear system. In the proposed ANFIS approach, the value of $x(k)$ that represents the static Hammerstein nonlinearity is distributed in the signals $\bar{w}_i$, produced in layer 3 (Fig. 5.1). Therefore, in coherence to the Hammerstein model, the linear system of the ANFIS scheme implements the adjustable IIR filter model exhibited in Eq. (5.15). Under such conditions, the functions f_i of the ANFIS approach are modeled as follows:

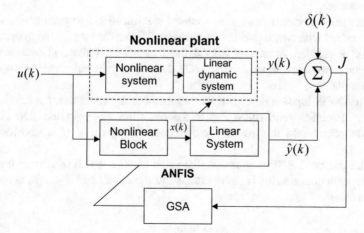

Fig. 5.3 Graphical representation of the identification process according to the proposed approach

Fig. 5.4 Structure of the ANFIS linear system

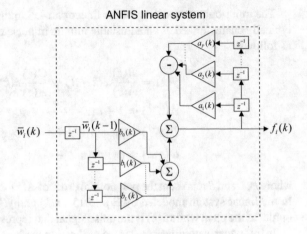

$$f_i(k) = \frac{B(z)}{A(z)} \bar{w}_i(k-1)$$

$$A(z^{-1}) = 1 + a_1 z^{-1} + \cdots + a_P z^{-P}$$

$$B(z^{-1}) = b_0 + b_1 z^{-1} + \cdots + b_L z^{-L-1}$$

$$f_i(k)A(z) = \bar{w}_i(k-1)B(z)$$

$$f_i(k)\left[1 + a_1 z^{-1} + \cdots + a_P z^{-P}\right] = \bar{w}_i(k-1)\left[b_0 + b_1 z^{-1} + a_L z^{-L-1}\right]$$

$$f_i(k) = b_0 \bar{w}_i(k-1) + b_1 \bar{w}_i(k-2) + \cdots + b_L \bar{w}_i(k-L-2)$$
$$- a_1 f_i(k-1) - \cdots - a_P f_i(k-P) \tag{5.16}$$

Therefore, each function f_i of the ANFIS scheme models an IIR filter which appropriately describes the dynamic behavior of the system within the fuzzy region specified by the antecedent of rule i. Figure 5.4 shows the structure of the ANFIS linear system.

Under these circumstances, the proposed identification approach consists of an ANFIS model with two inputs and one output. The inputs involve the system input $u(k)$ and a delayed version of $u(k)$ defined as $u(k\text{-}Q)$ (where Q represents the maximum considered input delay), while the output $y(k)$ represents the response of the model.

Each ANFIS input is described by two membership functions, considering two different Takagi-Sugeno rules. Figure 5.5 describes the proposed ANFIS identification scheme. In this paper, the maximum input delay Q is considered as 2.

In the proposed ANFIS approach, the value of $x(k)$, which represents the static Hammerstein nonlinearity, is distributed in the signals $\bar{w}_1$ and $\bar{w}_2$., $x(k)$ is modeled as follows:

$$x(k) = \bar{w}_1 + \bar{w}_2, \tag{5.17}$$

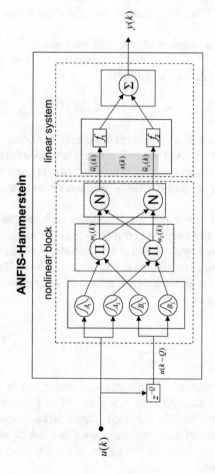

Fig. 5.5 The proposed ANFIS identification scheme

Due to the performed adaptation of the original ANFIS model in the linear system for each consequent function f_i, the number of linear parameters described in Sect. 2.2 has been altered. The new number of linear parameters P'_L is calculated as follows:

$$P'_L = 2 \cdot (P + L + 1); \tag{5.18}$$

where P and $L + 1$ represent the polynomial orders of the transference functions which implement the IIR filter. In this paper, each function is modeled considering $P = 3$ and $L = 2$.

Consequently, the total number of adaptive parameters P_T' in the proposed ANFIS-Hammerstein is $P_T' = P_{NL} + P'_L$. Therefore, the new parameter vector E'_{ANFIS} to be estimated is defined as follows:

$$E'_{ANFIS} = \{e_1^{ANFIS}, \ldots, e_{P'_T}^{ANFIS}\} \tag{5.19}$$

In our approach, the identification process consists of finding the optimal parameters of E'_{ANFIS} that present the best possible resemblance between the actual system and its Hammerstein alternative model. Therefore, the Hammerstein parameters represent the dimensions of each candidate solution (mass) for the identification problem.

To evaluate the quality of the identification under each parameter configuration (candidate solution), the mean squared error (MSE) criterion has been considered. The MSE index, symbolized by J, measures the similarity between the response $y(k)$ produced by actual plant and the estimation $\hat{y}(k)$ generated by its Hammerstein alternative model. The quality of each candidate solution (or parameter set) is evaluated according to the following model:

$$J = \frac{1}{NS} \sum_{k=1}^{NS} [y(k) - \hat{y}(k)]^2, \tag{5.20}$$

where NS expresses the number of input samples considered in the identification process. In this paper, to determine the complete set of parameters E'_{ANFIS}, the Gravitational Search Algorithm (GSA) is used. The objective function J is employed to evaluate the performance of each parameter set. The GSA algorithm evolves the group of candidate solutions based on the values of the objective function in order to find the optimal solution.

The value of Q, for our proposed approach has been set at $Q = 2$, in order to maintain compatibility with similar works reported in the literature [7–10, 46, 47], where the maximum considered delay is 2. Similarly, as they use the values of $P = 3$ and $L = 2$ to model the polynomial order of $A(z)$ and $B(z)$, in our approach we have used the same configuration.

5.4 Experimental Study

The main objective of this section is to present the modelling results of the proposed ANFIS-GSA method on numeric identification problems. In the test, a representative set of 7 experiments have been considered to evaluate the performance of the proposed method. From the configuration perspective, all the plants considered in the experiments hold the NARMAX structure (nonlinear autoregressive moving average model with exogenous inputs). However, according to their nonlinear characteristics, they are classified into the following categories:

(A) Hammerstein models with polynomial nonlinearities (Experiment 1 and Experiment 4). In a plant that maintains a polynomial nonlinearity, the relationship between the input variable and the output variable is modelled as an *n-th* degree polynomial in terms of the input. Hammerstein models with polynomial nonlinearity are difficult to identify, since algorithms must appropriately model the turning points of the plant [48]. Turning points are locations where the plant behavior significantly changes from sloping downwards to sloping upwards.

(B) Hammerstein models with nonlinearities modeled with linear piecewise functions (Experiment 2). Difficult nonlinear behaviors are frequently approximated through linear piecewise functions. However, the identification of Hammerstein models with linear piecewise functions represents a complex task, since its global model behavior must accurately approximate each of the single behaviors [49].

(C) Hammerstein models with trigonometric nonlinearities (Experiment 3 and Experiment 5). In a plant that maintains a trigonometric nonlinearity, the relationship between the input variable and the output variable is modelled as a trigonometric function of the input. The modelling of such systems under Hammerstein models is considered a difficult process, since trigonometric functions produce a limited output under high or low input values [50]. It is especially complex, since this nonlinearity cannot be proportionally modeled by the learning algorithms during the estimation.

(D) Hammerstein models from practical industrial problems (Experiment 6). Nonlinear plants that model practical engineering systems are usually complex, and their identification represents a challenge for any optimization method [51]. In this paper, we use the real, nonlinear dynamics of a heat exchanger system to identify a Hammerstein model.

All the ANFIS-GSA results obtained from the interactions listed above are also compared to those produced by the combination of the ANFIS architecture and other popular evolutionary algorithms. The final conclusions of the results have been verified by a statistical analysis of the experimental data.

In the experimental study, the proposed ANFIS-GSA approach has been used to identify the Hammerstein structure of the 7 plants while its results are compared to those produced by the DE method [52], the PSO method [53] and the original ANFIS algorithm [11]. The DE and PSO methods have been selected, since they obtain the highest number of references according to most popular scientific databases, ScienceDirect [54], SpringerLink [9] and IEEE-Xplore [55], over the last ten years. In the experiments, the population size N has been configured to 50 individuals. In order to eliminate the random effect, each experiment is tested for 30 independent runs. In the comparison, a fixed number of 1000 iterations has been considered as a stop criterion. This stop criterion has been chosen to keep compatibility with similar works published in the literature [7–10, 46, 47]. All experiments have been executed on a Pentium dual-core computer with 2.53-GHz and 8-GB RAM under MATLAB 8.3.

The methods used in the comparison have been set as follows:

1. **DE** [52]: The parameters are configured as $CR = 0.5$ and $F = 0.2$.
2. **PSO** [53]: The parameters are set to $c_1 = 2$ and $c_2 = 2$; The weight factor decreases linearly from 0.9 to 0.2.
3. **Original ANFIS** [11]: Two inputs, two membership functions ranging from $[-1,1]$ for experiments 1, 3, 4, and 5, and from $[-3,3]$ for experiment 2.
4. **GSA** [21]: Using the model $G_t = G_O e^{-\alpha \frac{t}{T}}$, where $\alpha = 20$, $G_O = 100$ and $T = 1000$.

These configurations have been selected because they achieve the best possible optimization performance according to their reported references [11, 21, 52, 53].

In the identification process, each algorithm obtains a solution to the parameter set E'_{ANFIS} at each execution. This estimation corresponds to the structure of the identified Hammerstein model with the lowest J value (Eq. 5.20). In the study, four performance indexes are used in the comparison: the worst identification value (W_J), the best identification value (B_J), the mean identification value (μ_J) and the standard deviation (σ_J). The first three indexes assess the accuracy of the identification between the actual plant and the estimated model whereas the latter evaluates the variability or robustness of the estimation. The worst identification value (W_J) expresses the least accurate estimation produced, which corresponds to the highest value of J (Eq. 5.20) obtained by each algorithm when considering a set of 30 distinct executions. The best identification value (B_J) indicates the most accurate model produced, which corresponds to the lowest value of J obtained in 30 executions. The mean identification value (μ_J) identifies the averaged estimation performance obtained by each algorithm in 30 independent runs. The Mean value simply represents the average value of J over the 30 different tests performed. The standard deviation (σ_J) is a measure that quantifies the dispersion of a set of the J data values. σ_J indicates the variability of the estimated models.

5.4.1   Experiment I

This identification problem has been extracted from [56, 57]. It considers the following model:

$$y(k) = \frac{B(z)}{A(z)}x(k-1) + \frac{C(z)}{A(z)}\delta(k), \qquad (5.21)$$

where

$$A(z) = 1 + 0.8z^{-1} + 0.6z^{-2},$$
$$B(z) = 0.4 + 0.2z^{-1},$$
$$C(z) = 1,$$
$$x(k) = u(k) + 0.5u^3(k) \qquad (5.22)$$

In the test, the values of the input signal $u(k)$ applied to the model Eq. (5.22) are random elements, uniformly distributed in the range $[-1,1]$. A noisy signal $\delta(k)$ is also injected as a white Gaussian sequence with a signal to noise ratio (SNR) of 70 dB. Under such conditions, the proposed ANFIS-GSA, the Differential Evolution (DE) [52], the Particle Swarm Optimization (PSO) method [53] and the original ANFIS algorithm [11] are used to identify the parameters E'_{ANFIS} of the model. All algorithms are randomly initialized in their first populations. The experimental results obtained from 30 independent executions are presented in Table 5.1. It reports the B_J, W_J and μ_J solutions in terms of J obtained during the 30 executions. The table also includes the σ_J values calculated in the 30 runs. According to Table 5.1, the proposed approach performs better than DE, PSO and the original ANFIS algorithm in terms of their μ_J values. This shows that the ANFIS-GSA method generally produces more accurate model estimations than the other evolutionary approaches DE and PSO, while the original ANFIS algorithm demonstrates the poorest performance. Table 5.1 also indicates that the ANFIS-GSA holds the best identification result (B_J) over 30 different executions. On the other hand, the W_J value generated by the original ANFIS algorithm represents superior performance compared to the other methods. With reference to the standard deviation index (σ_J), it is the original ANFIS method which presents the narrowest interval in comparison to the other evolutionary techniques. This case reflects the behavior of the gradient based techniques in multimodal optimization problems. Under such circumstances, the original ANFIS converges faster (lower σ_J values) but obtains a sub-optimal solution (higher μ_J values).

Figure 5.6 shows the ideal static nonlinearity $x(k)$ and the approximations produced by ANFIS-GSA, DE, PSO and the original ANFIS algorithm. From the

Table 5.1 Results of experiment I

		B_J	W_J	μ_J	σ_J
EX 1	DE	7.94E−04	2.80E−02	2.85E−03	3.08E−03
	PSO	1.13E−03	9.23E−02	2.19E−02	2.34E−02
	ANFIS	3.11E−03	5.73E−03	4.30E−03	5.10E−04
	ANFIS-GSA	2.92E−04	1.56E−02	1.03E−03	2.19E−03

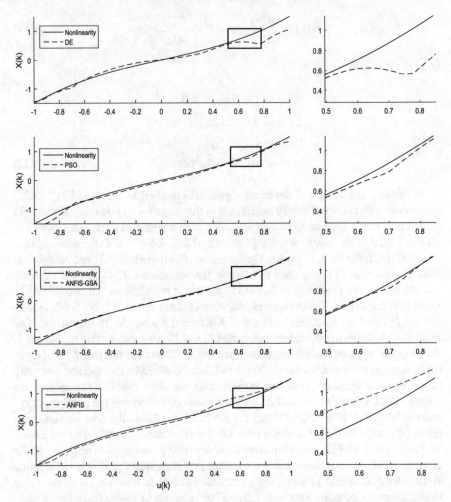

Fig. 5.6 Static nonlinearity $x(k)$ and the approximations produced by DE, PSO, the ANFIS-GSA and the original ANFIS for plant Eq. (5.22)

Fig. 5.6, it can be seen that the proposed ANFIS-GSA presents the best visual approximation among all methods. To better visualize the accuracy of each technique, Fig. 5.6 also incorporates a small zoom window which reproduces the system response with more detail.

In order to evaluate the performance of the identification process, the actual Hammerstein plant (Eq. 5.22) and the estimated models are tested with the same input signal $u(k)$ defined as follows:

$$u(k) = \begin{cases} \sin(\frac{2\pi k}{250}) & 0 < k \leqslant 500 \\ 0.8\sin(\frac{2\pi k}{250}) + 0.2\sin(\frac{2\pi k}{25}) & 500 < k \leqslant 700 \end{cases} \tag{5.23}$$

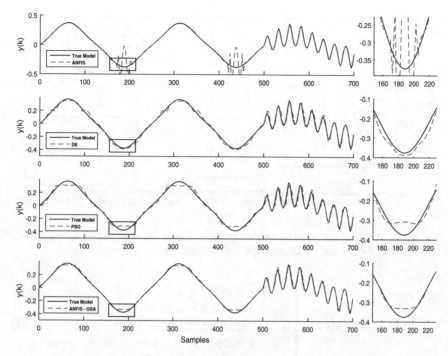

Fig. 5.7 Modeling results of the original ANFIS, DE, PSO, and the proposed ANFIS-GSA for Experiment I

Figure 5.7 shows the output $y(k)$ of each system. In the Figure, it is clear that each technique performs differently depending on the section of the system response. However, in general, the proposed ANFIS-GSA algorithm is the one that most resembles the true plant. Finally, Fig. 5.8 exhibits the convergence properties of the objective function J for the original ANFIS, DE, PSO and the proposed ANFIS-GSA. From the date in Fig. 5.8, we can be established that the proposed ANFIS-GSA reaches its final J values faster than its competitors.

5.4.2 Experiment II

The nonlinear plant considered in this test has been extracted from [46]. This model is similar to the model used in Experiment I, except that the input $u(k)$ and output of the static nonlinearity $x(k)$ in the system is modeled by a set of linear piecewise functions defined as follows:

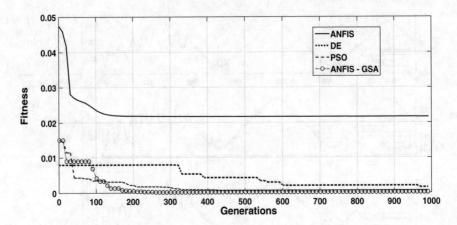

Fig. 5.8 Evolution of the objective function produced during the identification process by DE, PSO, the ANFIS-GSA, and the original ANFIS for Experiment I

$$x(k) = \begin{cases} -2.0 & \text{if } -3.0 \leqslant u(k) < -1.8 \\ \frac{u(k)}{0.6} + 1 & \text{if } -1.8 \leqslant u(k) < -0.6 \\ 0 & \text{if } -0.6 \leqslant u(k) < 0.6 \\ \frac{u(k)}{0.6} - 1 & \text{if } 0.6 \leqslant u(k) < 1.8 \\ 2 & \text{if } 1.8 \leqslant u(k) < 3.0 \end{cases}, \tag{5.24}$$

Similarly, in this experiment, the input $u(k)$ to the nonlinear plant as well as to the adaptive model is a random signal that is uniformly distributed in the range $[-3, 3]$. Under this setting, the ANFIS-GSA, DE, PSO and the original ANFIS algorithm are used to identify the parameters E'_{ANFIS} of the model. Table 5.2 presents the identification results after the parameter estimation. According to Table 5.2, the ANFIS-GSA method presents the best performance in terms of μ_J, B_J and W_J values. This means that the ANFIS-GSA algorithm produces better model estimations than the other methods with regard to the objective function J. The performance differences among ANFIS-GSA, DE and PSO can be attributed to their different capacities for handling multimodal objective functions. Although the original ANFIS reaches the minimal variability in its estimations (lower σ_J values), it converges most

Table 5.2 Results of experiment 2

		B_J	W_J	μ_J	σ_J
EX 2	DE	5.99E−03	7.67E−02	1.68E−02	9.57E−03
	PSO	6.78E−03	9.02E−02	3.54E−02	2.09E−02
	ANFIS	1.01E−02	1.04E−02	1.03E−02	4.23E−05
	ANFIS-GSA	2.02E−03	9.26E−03	4.59E−03	1.64E−03

of the times in a sub-optimal solution (higher μ_J values). Consequently, the proposed ANFIS-GSA holds the most accurate estimations with a competitive variability.

Figure 5.9 compares the nonlinearity $x(k)$ of the true Hammerstein system considered in this experiment and the nonlinearities produced by the four algorithms after the training. Figure 5.9 shows how difficult it is to accurately approximate the set of linear piecewise functions with the estimated models. Even so, the proposed ANFIS-GSA algorithm obtains the best visual approximation in comparison to the other techniques. Figure 5.10 shows the output value of each method when using the input value defined in Eq. (5.23). This demonstrates the remarkable nonlinear identification capabilities of the ANFIS model trained by using GSA. In the Figure, it is clear that the proposed ANFIS-GSA method obtains the best identification results compared

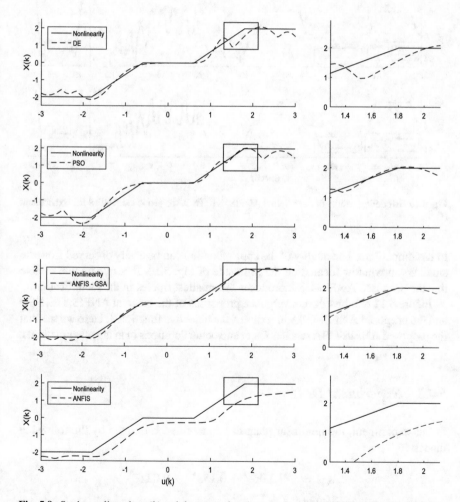

Fig. 5.9 Static nonlinearity $x(k)$ and the approximations produced by DE, PSO, the ANFIS-GSA, and the original ANFIS for pant Eqs. (5.22) and (5.24)

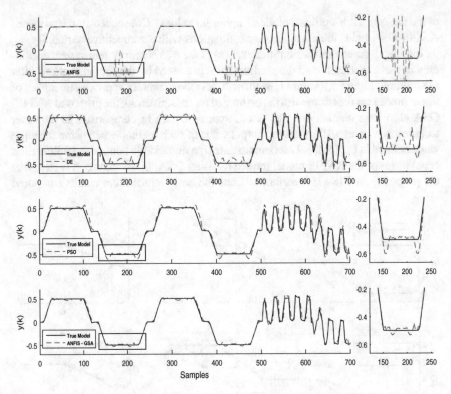

Fig. 5.10 Modeling results of the original ANFIS, DE, PSO, the proposed ANFIS for Experiment II

to its competitors. The quality of the approximation can be easily observed from the small zoom window located on the right side of Fig. 5.10. It is remarkable to note that the proposed ANFIS-GSA produces the smallest ripples in the identification.

Figure 5.11 exhibits the convergence properties of the original ANFIS, DE, PSO and the proposed ANFIS-GSA in terms of the objective function J. Here we see that the proposed ANFIS-GSA reaches faster and better solutions than the other methods.

5.4.3 Experiment III

In this experiment, the nonlinear plant to be identified is defined by the following model:

$$A(z) = 1 + 0.9z^{-1} + 0.15z^{-2} + 0.02z^{-3},$$
$$B(z) = 0.7 + 1.5z^{-1},$$
$$C(z) = 1,$$

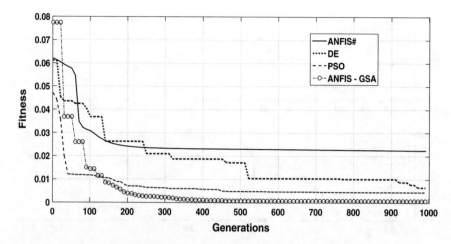

Fig. 5.11 Evolution of the objective function produced during the identification process by DE, PSO, the ANFIS-GSA and the original ANFIS for Experiment II

$$x(k) = 0.5\sin^3(\pi u(k)) \tag{5.25}$$

This model has been extracted from [47]. In the experiment, the input signal $u(k)$ used in the identification is defined as random numbers uniformly distributed in the range $[-1,1]$. Under these circumstances, the experimental data are used for the proposed ANFIS-GSA, DE, PSO and the original ANFIS algorithm to identify the plant through the determination of the parameter vector E'_{ANFIS}. Table 5.3 presents the identification results after the parameter estimation. Figure 5.12 shows the original nonlinearity as well as the nonlinearities estimated after the modelling using the four algorithms. Figure 5.13 compares outputs of the actual system to the model outputs after identification, considering the input $u(k)$ defined in Eq. (5.23). Figure 5.14 exhibits the convergence values of the objective function J as the identification process evolves. An analysis of the experimental data demonstrates that the original ANFIS and the proposed ANFIS-GSA perform better than the other methods.

A detailed inspection of Table 5.3 shows that the original ANFIS produces a better result than the proposed ANFIS-GSA in terms of accuracy. This fact could be attributed to the nonlinear characteristics of the plant being modelled (Eq. 5.25).

Table 5.3 Results of experiment 3

		B_J	W_J	μ_J	σ_J
EX 3	DE	7.01E−03	2.21E−02	1.18E−02	2.81E−03
	PSO	1.06E−02	9.94E−02	3.81E−02	2.17E−02
	ANFIS	5.66E−03	2.84E−01	4.01E−02	1.51E−06
	ANFIS-GSA	5.82E−03	3.59E−01	1.07E−02	4.92E−03

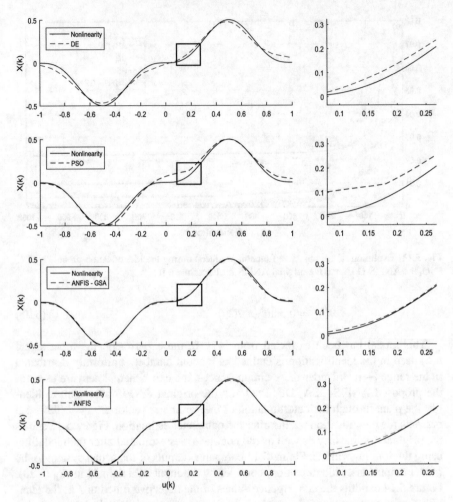

Fig. 5.12 Static nonlinearity $x(k)$ and the approximations produced by DE, PSO, the ANFIS-GSA, and the original ANFIS for pant Eq. (5.25)

Since its nonlinear behavior (Fig. 5.12) can be approximated by a linear mixture of Gaussians, an objective function J with only one minimum is produced. It is well known that gradient based techniques, such as the back-propagation used by the original ANFIS, are better than evolutionary methods to find the optimal solution with regard to the accuracy and number of iterations in such cases [58].

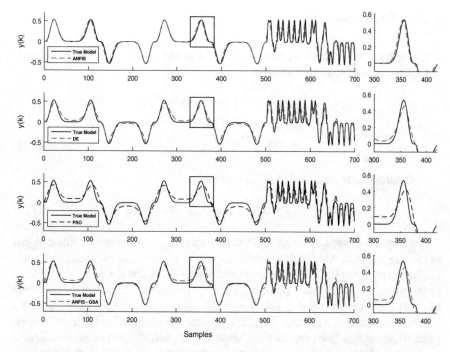

Fig. 5.13 Modeling results of the original ANFIS, DE, PSO, the proposed ANFIS for Experiment III

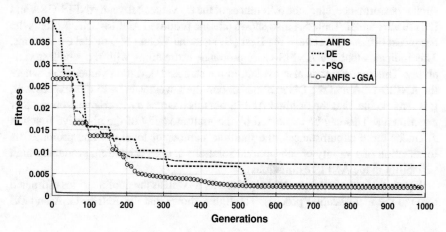

Fig. 5.14 Evolution of the objective function produced during the identification process by DE, PSO, the ANFIS-GSA and the original ANFIS for Experiment III

5.4.4   Experiment IV

In this experiment, the transfer functions $A(z)$, $B(z)$ and $C(z)$ that model the nonlinear plant maintain the following structure:

$$A(z) = 1 - 1.5z^{-1} + 0.7z^{-2},$$
$$B(z) = z^{-1} + 0.5z^{-2},$$
$$C(z) = 1 - 0.5z^{-1}, \qquad (5.26)$$

The nonlinearity proposed in [47] is defined as:

$$x(k) = u(k) + 0.5u^2(k) + 0.3u^3(k) + 0.1u^4(k) \qquad (5.27)$$

The input $u(k)$ used in the identification is a uniformly distributed random signal in the range $[-1,1]$. Similar to the other experiments, the parameter vector E'_{ANFIS} is updated by using PSO, DE, the proposed ANFIS-GSA and the original ANFIS in order to minimize J. In this experiment, since the polynomial degree of the nonlinear block is high, several turning points must be approximated. As a result, a multimodal surface in J is expected. Table 5.4 presents the results after the identification process. According to the results from Table 5.4, the ANFIS-GSA obtains the lowest μ_J value, indicating that, in general, this method produces the best model estimations. On the other hand, the estimations generated by the DE approach maintain a similar performance, since the differences of the μ_J values between ANFIS-GSA and DE are very small. Table 5.4 also shows that the proposed ANFIS-GSA presents the narrowest difference between the Best (B_J) and the Worst (W_J) model estimations. This indicates that the ANFIS-GSA produces estimations with the smallest variability. This robustness is also evident in the standard deviation values (σ_J), where the ANFIS-GSA and the DE methods obtain the lowest values. In this experiment, it is remarkable, that the original ANFIS obtains a worse σ_J in comparison to other experiments. This can be explained by the multimodality of the objective function J. Under these circumstances, the multiple number of local minima promote the localization of a variety of sub-optimal solutions at each execution, producing high variability in the ANFIS estimations.

Figure 5.15 presents the actual nonlinearity as well as the nonlinearities estimated through the identification process. The figure shows that the ANFIS-GSA and DE

Table 5.4 Results of Experiment 4

		B_J	W_J	μ_J	σ_J
EX 4	DE	2.02E−03	1.42E−02	3.58E−03	1.36E−03
	PSO	2.08E−03	8.98E−02	1.38E−02	1.69E−02
	ANFIS	6.50E−03	7.44E−01	7.75E−03	6.18E−02
	ANFIS-GSA	1.65E−03	5.21E−03	2.48E−03	9.01E−04

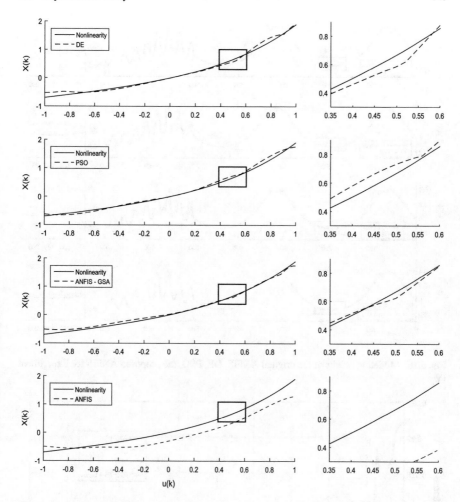

Fig. 5.15 Static nonlinearity *x(k)* and the approximations produced by DE, PSO, the ANFIS-GSA, and the original ANFIS for pant Eqs. (5.26) and (5.27)

present the best visual approximations of the actual nonlinearity. Furthermore, the comparison of the actual plant and the identified models are shown in Fig. 5.16, using the input $u(k)$ defined in Eq. (5.23). Figure 5.16 demonstrates that the approximation quality of the ANFIS-GSA is superior to the other algorithms. This can easily be seen from the small zoom windows shown on the right side of the Figure. Finally, Fig. 5.17 exhibits the convergence values of the objective function J as the identification process evolves. After an analysis of Fig. 5.17, it is clear that the proposed ANFIS-GSA produces convergence properties superior to the other methods.

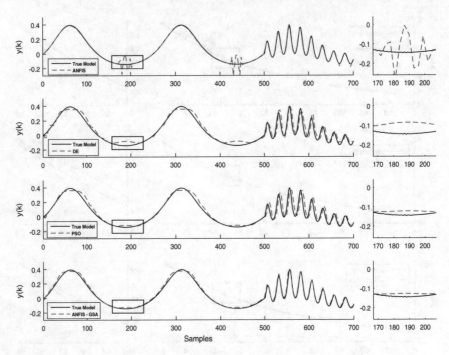

Fig. 5.16 Modeling results of the original ANFIS, DE, PSO, the proposed ANFIS for Experiment IV

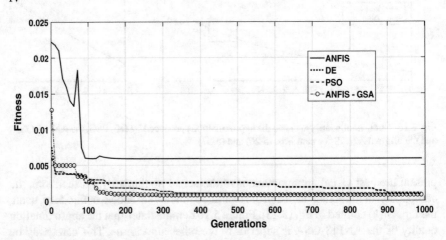

Fig. 5.17 Evolution of the objective function produced during the identification process by DE, PSO, the ANFIS-GSA and the original ANFIS for Experiment IV

5.4.5 *Experiment V*

In this experiment, we consider the plant taken from [9] which is described as follows:

$$A(z) = 1 + 0.3z^{-1} + 0.65z^{-2} - 0.35z^{-3},$$
$$B(z) = 0.125 - 0.6z^{-1},$$
$$C(z) = 1,$$

$$x(k) = 0.5\sin^3(\pi u(k)) - \frac{2}{u^3(k) + 2} - 0.1\cos(4\pi u(k)) + 1.125 \qquad (5.28)$$

In this test, the input signal $u(k)$ applied to the nonlinear plant is obtained as a random number uniformly distributed in the range $[-1,1]$, which produces a set of experimental data. With the data, the proposed ANFIS-GSA, DE, PSO and the original ANFIS algorithm are used to identify the parameters E'_{ANFIS} of the model.

In this experiment, since the trigonometric expression of the nonlinear block involves a complex structure, a high multimodal objective function J is produced. The experimental results obtained from 30 independent executions are presented in Table 5.5. According to Table 5.5, the proposed approach performs better than DE, PSO and the original ANFIS algorithm in terms of B_J, W_J, μ_J and Standard Deviation Values. The performance differences among ANFIS-GSA, DE and PSO can be attributed to their different capacities for handling multimodal objective functions. In this test, similar to Experiment IV, the original ANFIS obtains a high σ_J value due to the multimodality of the objective function J. Figure 5.18 shows the ideal static nonlinearity $x(k)$ and the approximations produced by ANFIS-GSA, DE, PSO and the original ANFIS algorithm. Figure 5.18 shows that the proposed ANFIS-GSA method produces the best visual similarity to the actual nonlinearity.

To validate the quality of the identification, the input signal $u(k)$, defined in Eq. (5.29), has been used as signal test.

$$u(k) = \begin{cases} \sin(\frac{2\pi k}{250}) & 0 < k \leq 250 \\ 0.8\sin(\frac{2\pi k}{250}) + 0.2\sin(\frac{2\pi k}{25}) & 250 < k \leq 800 \end{cases} \qquad (5.29)$$

In Fig. 5.19, the performance comparison of the actual plant and the identified models are shown. From the Figure, it is clear that the proposed ANFIS-GSA reaches

Table 5.5 Results of Experiment 5

		B_J	W_J	μ_J	σ_J
EX 5	DE	8.53E−04	9.34E−03	2.28E−03	1.05E−03
	PSO	1.33E−03	9.71E−02	1.29E−02	1.81E−02
	ANFIS	3.82E−03	5.58E−03	4.68E-03	3.80E−04
	ANFIS-GSA	3.79E−04	5.05E−03	1.16E-03	1.01E−04

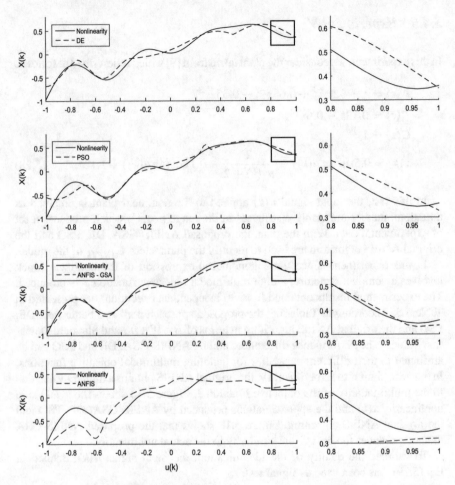

Fig. 5.18 Static nonlinearity $x(k)$ and the approximations produced by DE, PSO, the ANFIS-GSA, and the original ANFIS for pant Eq. 5.28

the best visual approximation among all approaches. Figure 5.20 contains the convergence properties of objective function J for the original ANFIS, DE, PSO and the proposed ANFIS-GSA. An analysis of the convergence results demonstrates that the proposed ANFIS-GSA obtains the best performance among all methods.

5.4.6 Experiment VI

The nonlinear plant considered for this example is a model that includes a trigonometric nonlinearity extracted from [60] with the following definition:

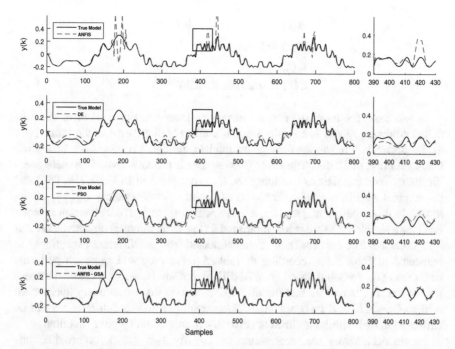

Fig. 5.19 Modeling results of the original ANFIS, DE, PSO, the proposed ANFIS for Experiment V

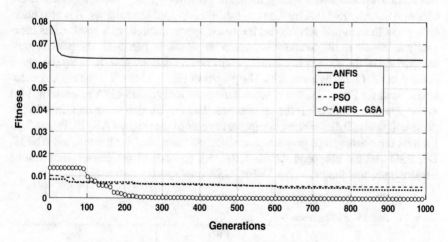

Fig. 5.20 Evolution of the objective function produced during the identification process by DE, PSO, the ANFIS-GSA and the original ANFIS for Experiment V

$$A(z) = 1 + 0.9z^{-1} + 0.7z^{-2},$$
$$B(z) = 0.4z^{-1} - 0.5z^{-2},$$
$$C(z) = 1,$$
$$x(k) = u(k) \cdot \cos(\pi u(k)) \qquad (5.30)$$

In this case, the input $u(k)$ to the exact nonlinear plant as well as to the estimated nonlinear model is a random signal uniformly distributed in the range $[-2,
2]$. Therefore, before passing through the models, the signal is assumed to be a white
Gaussian noise of 50 dB., The set of training data is produced based on these spec-
ifications. With this data as a training set, the proposed ANFIS-GSA, DE, PSO and
the original ANFIS algorithm are used to identify the parameters E'_{ANFIS} of the
model. In this experiment, since the trigonometric expression of the nonlinear block
involves a simple structure, a low multimodal (few local minima) objective function
J is produced. The experimental results collected from 30 different executions are
registered in Table 5.6. According to Table 5.6, the proposed approach performs
better than DE, PSO and the original ANFIS algorithm. However, a detailed inspection of the results reveals that the differences among the evolutionary approaches
ANFIS-GSA, DE and PSO are small. Such small performance differences can be
attributed to their similar abilities to solve low multimodal objective functions.

Figure 5.21 shows the ideal static nonlinearity $x(k)$ and the approximations
produced by ANFIS-GSA, DE, PSO and the original ANFIS algorithm. Figure 5.21
shows that all the evolutionary methods present a good visual approximation,
although the ANFIS-GSA method has the best perceptible similarity. It is also interesting that the original ANFIS has the worst approximation, as a result of its inca-
pacity to handle multimodal optimization problems. In Fig. 5.22, the performance
comparison of the actual plant and the identified models are shown, using the signal
defined in Eq. (5.23) as input $u(k)$. The graphs of Fig. 5.22 confirm the same conclusions found for Fig. 5.21. It is clear that the proposed ANFIS-GSA reaches the best
visual approximation of all the approaches. The graphs also show that they do not
present high visual differences, except for the case of the original ANFIS. Figure 5.23
exhibits the convergence properties of objective function J for the original ANFIS,
DE, PSO and the proposed ANFIS-GSA. An analysis of the convergence results
demonstrates that the proposed ANFIS-GSA produces the best results.

Table 5.6 Results of Experiment 6

		B_J	W_J	μ_J	σ_J
EX 6	DE	1.40E−03	3.98E−02	8.10E−03	5.60E−03
	PSO	7.22E−03	4.29E−02	5.60E−03	6.90E−03
	ANFIS	5.30E−03	5.60E−01	12.50E−03	7.32E−05
	ANFIS-GSA	5.14E−−04	4.50E−03	2.30E−03	1.20E−03

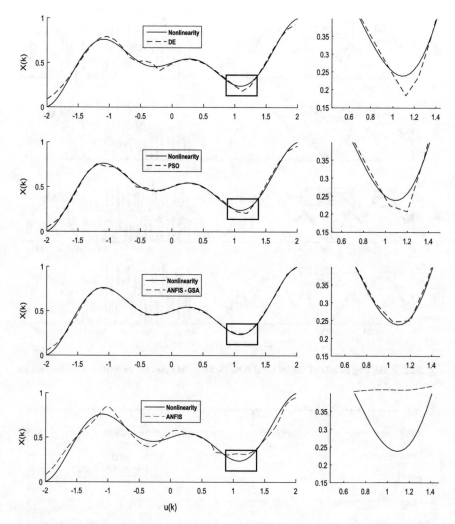

Fig. 5.21 Static nonlinearity $x(k)$ and the approximations produced by DE, PSO, the ANFIS-GSA and the original ANFIS for plant Eq. (5.30)

5.4.7 Experiment VII

The nonlinear Hammerstein plant considered in this experiment has been proposed to model the real nonlinear dynamics of a heat exchanger system. This practical industrial problem has been extracted from [51]. Therefore, the nonlinear dynamics of this system are modeled according to the following equations:

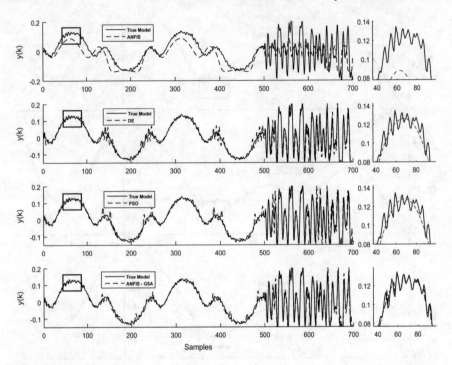

Fig. 5.22 Modeling results of the original ANFIS, DE, PSO and the proposed ANFIS-GSA for Experiment VI

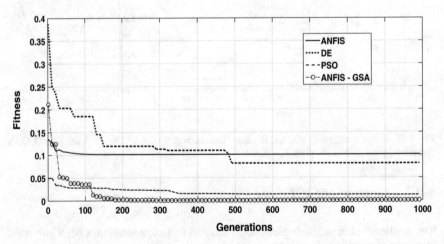

Fig. 5.23 Evolution of the objective function produced during the identification process by DE, PSO, the ANFIS-GSA and the original ANFIS for Experiment VI

$$A(z) = 1 - 1.608z^{-1} + 0.6385z^{-2},$$
$$B(z) = 0.207z^{-1} - 0.1764z^{-2},$$
$$C(z) = 1,$$
$$x(k) = -31.549x(n) + 41.732x^2(n) - 24.201x^3(n) + 68.634x^4(n) \qquad (5.31)$$

As can be seen, the main characteristics of this plant are similar to those present in Experiment 4. Like other tests, in this experiment the input to the nonlinear plant and to the adaptive model is a random signal that is uniformly distributed in the range $[-1, 1]$. Under these settings, the ANFIS-GSA, DE, PSO and the original ANFIS algorithm are used to identify the parameters $\mathbf{E}'_{\text{ANFIS}}$ of the model. Table 5.7 reports their results in the optimization of the objective function J (Eq. 5.20). Table 5.7 clearly shows that the proposed ANFIS-GSA performs best in terms of the B_J, W_J, μ_J and σ_J values. This indicates that the ANFIS-GSA algorithm obtains the most accurate model estimations for the plant compared to the other methods. Due to the multimodality, the original ANFIS maintains the worst performance out of the complete set of techniques.

Figure 5.24 shows the comparison between the true nonlinearity produced by the Hammerstein model and the estimations generated by the ANFIS-GSA, DE, and PSO methods and the original ANFIS. According to the Figure, the ANFIS-GSA produces the best visual approximations of the actual nonlinearity. However, the other evolutionary techniques such as PSO and DE offer a competitive approximation. The exception is the original ANFIS which produces the worst approximation due to its difficulties to operate with multimodal optimization formulations.

In Fig. 5.25, the performance comparison between the actual plant and the identified models is shown. The responses consider the signal defined in Eq. (5.23) as the input. The graphs of Fig. 5.25 confirm the same conclusions as those found in Fig. 5.24. From the Figure, it is clear that the proposed ANFIS-GSA generates a better visual approximation the other approaches. Figure 5.26 exhibits the convergence properties of objective function J for the original ANFIS, DE, PSO and the proposed ANFIS-GSA. The convergence results demonstrate that the proposed ANFIS-GSA performs better than the other methods tested.

Table 5.7 Results of Experiment 7

		B_J	W_J	μ_J	σ_J
EX 7	DE	2.74E−02	1.30E+00	4.25E−01	3.11E−01
	PSO	2.90E−03	3.13E−01	6.49E−02	1.55E−01
	ANFIS	1.54E+00	1.54E+00	1.54E+00	3.14E−04
	ANFIS-GSA	2.62E−03	6.74E−02	7.63E−03	7.03E−03

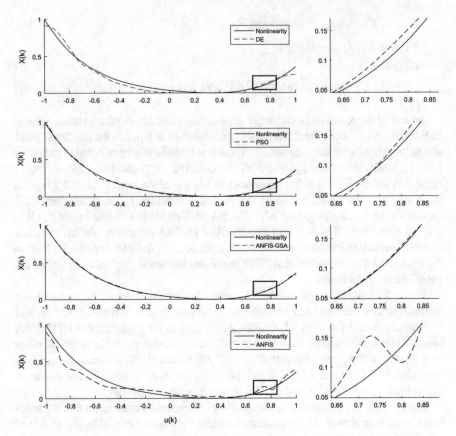

Fig. 5.24 Static nonlinearity $x(k)$ and the approximations produced by DE, PSO, the ANFIS-GSA and the original ANFIS for plant Eq. (5.31)

5.4.8   Statistical Analysis

To statistically analyze the results of Tables 5.1, 5.2, 5.3, 5.4. 5.4, 5.6 and 5.7, a non-parametric test known as the Wilcoxon analysis [61,62] has been conducted. It allows us to evaluate the differences between two related methods. The test is performed for the 5% (0.005) significance level over the μ_J data produced in each experiment. Table 5.8 reports the p-values generated by the Wilcoxon analysis for the pair-wise comparison among the algorithms. For the test, three groups are created: ANFIS-GSA vs. DE, ANFIS-GSA vs. PSO and ANFIS-GSA vs. ANFIS. In the Wilcoxon analysis, it is considered a null hypothesis if there is no a notable difference between the two methods. On the other hand, an alternative hypothesis is admitted if there is an important difference between the approaches. In order to facilitate the analysis of Table 5.8, the symbols ▲, ▼, and ▶ are adopted. ▲ indicates that the proposed method performs significantly better than the tested algorithm on the

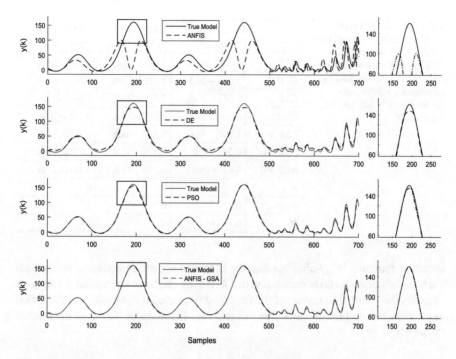

Fig. 5.25 Modeling results of the original ANFIS, DE, PSO and the proposed ANFIS-GSA for Experiment VII

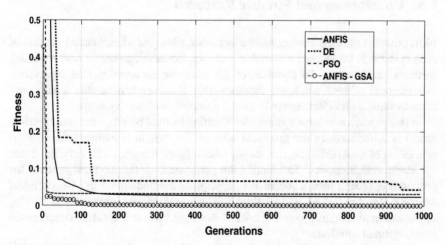

Fig. 5.26 Evolution of the objective function produced during the identification process by DE, PSO, the ANFIS-GSA and the original ANFIS for Experiment VII

Table 5.8 p-values produced by Wilcoxon's test comparing ANFIS-GSA versus DE, ANFIS-GSA versus PSO and ANFIS-GSA versus ANFIS over the μ_J values from Tables 5.1, 5.2, 5.3, 5.4. 5.4, 5.6 and 5.7

	ANFIS-GSA versus DE	ANFIS-GSA versus PSO	ANFIS-GSA versus ANFIS
EXP I	1.28E−22▲	2.60E−22▲	1.16E−23▲
EXP II	2.64E−24▲	1.56E−30▲	2.54E−28▲
EXP III	2.56E−14▲	1.51E−34▲	1.54E−01►
EXP IV	2.60E−10▲	8.89E−21▲	6.74E−10▲
EXP V	1.20E−10▲	7.88E−08▲	6.74E−12▲
EXP VI	4.20E−4▲	3.17E−05▲	1.07E−10▲
EXP VII	6.45E−5▲	2.29E−05▲	3.11E−11▲
▲	7	7	6
►	0	0	1
▼	0	0	0

specified function. ▼ symbolizes that the proposed algorithm performs worse than the tested algorithm, and ► means that the Wilcoxon rank sum test cannot distinguish between the simulation results of the proposed multimodal optimizer and the tested algorithm. The number of cases that fall in these situations are shown at the bottom of the table.

5.5 Conclusions and Further Research

Most practical systems have nonlinear behavior. Since the identification process of such systems is considered a complex problem, the development of new modeling methods has attracted the attention of the scientific community. One representative example of such modeling methods uses Hammerstein models which have demonstrated their effectiveness to model complex nonlinear systems.

In this paper, a nonlinear system identification method based on the Hammerstein model is introduced. In the proposed scheme, the system is modeled through the adaptation of an ANFIS scheme, taking advantage of the similitude between it and the Hammerstein model. To identify the parameters of the modeled system, the proposed approach uses a recent nature inspired method called the Gravitational Search Algorithm (GSA)., GSA solves complex problems better than most existing optimization algorithms, avoiding critical flaws such as the premature convergence to sub-optimal solutions.

To demonstrate the ability and robustness of our approach, the proposed identification system has been experimentally evaluated over a test suite of 7 complex experiments. To assess the performance of the proposed approach, it has been compared to other popular evolutionary approaches such as the Particle Swarm Optimization (PSO), Differential Evolution (DE) and the original ANFIS training algorithm. The results, statistically validated by the Wilcoxon analysis, have confirmed that the proposed algorithm outperforms other evolutionary algorithms for most of the test experiments in terms of its solution quality.

As this paper has shown, Hammerstein modeling trough evolutionary computation methods offers a number of interesting topics on which further research would be beneficial. Further directions include: (1) The definition of new Hammerstein benchmark plants that allow the appropriate evaluation of identification techniques. Currently, the set of benchmark models considers the most typical nonlinearities, for this reason, the integration of other complexities already known in traditional nonlinear plants to Hammerstein models is important. (2) The consideration of multi-objective optimization techniques in order to improve the accuracy modeling of both nonlinear and linear blocks. The approach proposed in this paper considers the identification process as an optimization problem with only one objective (minimize Eq. 5.20). The optimization model can be extended to consider two different objectives. The first objective involves the approximation of the nonlinear block while the second objective expresses the quality of the identification of the linear filter. (3) The construction of a new mono-objective function that involves structural elements in order to optimize not only the ANFIS parameters, but also its architecture. The objective function proposed in this work could be modified to add the complete or partial structure of the ANFIS system. The idea is to modify adaptability in the optimization process of ANFIS architecture so that the final optimized ANFIS structure only includes the necessary elements for accurately modeling the Hammerstein system at hand.

References

1. Piroddi, L., Farina, M., Lovera, M.: Black box model identification of nonlinear input–output models: a Wiener-Hammerstein benchmark. Control Eng. Pract. **20**(11), 1109–1118 (2012)
2. Huo, H., Zhu, X., Hu, W., Tu, H., Li, J., Yang, J.: Nonlinear model predictive control of SOFC based on a Hammerstein model. J. Power Sources **185**(1), 38–344 (2008)
3. Jurado, F.: A method for the identification of solid oxide fuel cells using a Hammerstein model. J. Power Sources **154**(1), 145–152 (2006)
4. Gilabert, P., Montoro, G., Bertran, E.: On the wiener and Hammerstein models for power amplifier predistortion, Asia-Pacific Microw. Conf. Proceedings, APMC. **2**(2), 5–8 (2005)
5. Smith, J.G., Kamat, S., Madhavan, K.P.: Modeling of pH process using wavenet based Hammerstein model. J. Process Control **17**(6), 551–561 (2007)
6. Zhang, X., Tan, Y.: Modelling of ultrasonic motor with dead-zone based on Hammerstein model structure. J. Zhejiang Univ. Sci. A **9**(1), 58–64 (2005)
7. Cui, M., Liu, H., Li, Z., Tang, Y., Guan, X.: Identification of Hammerstein model using functional link artificial neural network. Neurocomputing **142**, 419–428 (2014)
8. Tang, Y., Li, Z., Guan, X.: Identification of nonlinear system using extreme learning machine based Hammerstein model. Commun. Nonlinear Sci. Numer. Simul. **19**(9), 3171–3183 (2014)
9. Nanda, S.J., Panda, G., Majhi, B.: Improved identification of Hammerstein plants using new CPSO and IPSO algorithms. Expert Syst. Appl. **37**(10), 6818–6831 (2010)
10. Gotmare, A., Patidar, R., George, N.V.: Nonlinear system identification using a cuckoo search optimized adaptive Hammerstein model. Expert Syst. Appl. **42**(5), 2538–2546 (2015)
11. Jang, J.S.R.: ANFIS: adaptive-network-based fuzzy inference system. IEEE Trans. Syst. Man Cybern. **23**(3), 665–685 (1993)

12. Takagi, T., Sugeno, M.: Fuzzy identification of systems and its applications to modeling and control. IEEE Trans. Syst. Man Cybern. SMC **15**, 116–132 (1985)
13. Vairappan, C., Tamura, H., Gao, S., Tang, Z.: Batch type local search-based adaptive neuro-fuzzy inference system (ANFIS) with self-feedbacks for time-series prediction. Neurocomputing **72**(7–9), 1870–1877 (2009)
14. Mohandes, M., Rehman, S., Rahman, S.M.: Estimation of wind speed profile using adaptive neuro-fuzzy inference system (ANFIS). Appl. Energy **88**(11), 4024–4032 (2011)
15. Salahshoor, K., Kordestani, M., Khoshro, M.S.: Fault detection and diagnosis of an industrial steam turbine using fusion of SVM (support vector machine) and ANFIS (adaptive neuro-fuzzy inference system) classifiers. Energy **35**(12), 5472–5482 (2010)
16. Engin, S.N., Kuvulmaz, J., Ömurlü, V.E.: Fuzzy control of an ANFIS model representing a nonlinear liquid-level system. Neural Comput. Appl. **13**(3), 202–210 (2004)
17. Wu, X.J., Zhu, X.J., Cao, G.Y., Tu, H.Y.: Nonlinear modeling of a SOFC stack based on ANFIS identification. Simul. Model. Pract. Theory **16**(4), 399–409 (2008)
18. Zelinka, I.: A survey on evolutionary algorithms dynamics and its complexity—mutual relations, past, present and future. Swarm Evolut. Comput. **25**, 2–14 (2015)
19. Al-Duwaish, H.N.: Nonlinearity structure using particle swarm optimization. Arabian J. Sci. Eng. **36**(7), 1269–1276 (2011)
20. Mete, S., Ozer, S., Zorlu, H.: System identification using Hammerstein model optimized with differential evolution algorithm. AEU Int. J. Electron. Commun. **70**(12), 1667–1675 (2016)
21. Rashedi, E., Nezamabadi-pour, H., Saryazdi, S.: GSA: a gravitational search algorithm. Inf. Sci. **179**(13), 2232–2248 (2009)
22. Farivar, F., Shoorehdeli, M.A.: Stability analysis of particle dynamics in gravitational search optimization algorithm. Inf. Sci. **337–338**, 25–43 (2016)
23. Yazdani, S., Nezamabadi-Pour, H., Kamyab, S.: A gravitational search algorithm for multimodal optimization. Swarm Evol. Comput. **14**, 1–14 (2014)
24. Yazdani, A., Jayabarathi, T., Ramesh, V., Raghunathan, T.: Combined heat and power economic dispatch problem using firefly algorithm. Front. Energy **7**(2), 133–139 (2013)
25. Kumar, V., Chhabra, J.K., Kumar, D.: Automatic cluster evolution using gravitational search algorithm and its application on image segmentation. Eng. Appl. Artif. Intell. **29**, 93–103 (2014)
26. Zhang, W., Niu, P., Li, G., Li, P.: Forecasting of turbine heat rate with online least squares support vector machine based on gravitational search algorithm. Knowledge-Based Syst. **39**, 34–44 (2015)
27. Karaboga, D., Kaya, E.: An adaptive and hybrid artificial bee colony algorithm (aABC) for ANFIS training. Appl. Soft Comput. **49**, 423–436 (2016)
28. Sarkheyli, A., Zain, A.M., Sharif, S.: Robust optimization of ANFIS based on a new modified GA. Neurocomputing **166**, 357–366 (2015)
29. Wei, L.Y.: A GA-weighted ANFIS model based on multiple stock marketvolatility causality for TAIEX forecasting. Appl. Soft Comput. **13**(2), 911–920 (2013)
30. Palupi Rini, D., Mariyam Shamsuddin, S., Sophiayati Yuhaniz, S.: Particle swarm optimization for ANFIS interpretability and accuracy, Soft Comput. **20**(1), 251–262 (2016)
31. Wang, J.S., Ning, C.X.: ANFIS based time series prediction method of bank cash flow optimized by adaptive population activity PSO algorithm. Information **6**(3), 300–313 (2015)
32. Liu, P., Leng, W., Fang, W.: Training ANFIS model with an improved quantum-behaved particle swarm optimization algorithm, Math. Probl. Eng. (2013)
33. Catalão, J.P.S., Pousinho, H.M.I., Mendes, V.M.F.: Hybrid Wavelet-PSO-ANFIS approach for short-term electricity prices forecasting. IEEE Trans. Power Syst. **26**(1), 137–144 (2011)
34. Suja Priyadharsini, S., Edward Rajan, S., Femilin Sheniha, S.: A novel approach for the elimination of artefacts from EEG signals employing an improved artificial immune system algorithm, J. Exp. Theor. Artif. Intell. **21**, 1–21 (2015)
35. Gunasekaran, M., Ramaswami, K.S.: A fusion model integrating ANFIS and artificial immune algorithm for forecasting indian stock market. J. Appl. Sci. **11**(16), 3028–3033 (2011)
36. Asadollahi-Baboli, M.: In silico prediction of the aniline derivatives toxicitiesto Tetrahymena pyriformis using chemometrics tools. Toxicol. Environ. Chem. **94**(10), 2019–2034 (2012)

37. Teimouri, R., Baseri, H.: Optimization of magnetic field assisted EDM using the continuous ACO algorithm. Appl. Soft Comput. **14**, 381–389 (2014)
38. Varshini, G.S., Raja, S.C., Venkatesh, P.: Design of ANFIS controller for power system stability enhancement using FACTS device. In: Power Electronics and Renewable Energy Systems, pp. 1163–1171. Springer, India (2015)
39. Bhavani, M., Selvi, K., Sindhumathi, L.: Neuro fuzzy load frequency control in a competitive electricity market using BFOA tuned SMES and TCPS. In: Swarm, evolutionary, and memetic computing, pp. 373–385. Springer International Publishing (2014)
40. Azarbad, M., Azami, H., Sanei, S., Ebrahimzadeh, A.: New neural network-based approaches for GPS GDOP classification based on neuro-fuzzy inference system, radial basis function, and improved bee algorithm. Appl. Soft Comput. **25**, 285–292 (2014)
41. Rani, M.A.F., Sankaragomathi, B.: Performance enhancement of PID controllers by modern optimization techniques for speed control of PMBL DC motor. Res. J. Appl. Sci. Eng. Technol. **10**(10), 1154–1163 (2015)
42. Chen, J.F., Do, Q.H.: A cooperative cuckoo search–hierarchical adaptive neuro-fuzzy inference system approach for predicting student academic performance. J. Intell. Fuzzy Syst. **27**(5), 2551–2561 (2014)
43. Khazraee, S.M., Jahanmiri, A.H., Ghorayshi, S.A.: Model reduction and optimization of reactive batch distillation based on the adaptive neuro-fuzzy inference system and differential evolution. Neural Comput. Appl. **20**(2), 239–248 (2011)
44. Zangeneh, A.Z., Mansouri, M., Teshnehlab, M., Sedigh, A.K.: Training ANFIS system with DE algorithm. In: Advanced Computational Intelligence (IWACI),2011 Fourth International Workshop on IEEE, pp. 308–314 (2011)
45. Afifi, F., Badrul Anuar, N., Shamshirband, S., Raymond, K.K.: Choo, DyHAP: dynamic hybrid ANFIS-PSO approach for predicting mobile malware. Plose one (2016). https://doi.org/10.1371/journal.pone.0162627
46. Nanda, S.J., Panda, G., Majhi, B.: Improved identification of Hammerstein plants using new CPSO and IPSO algorithms. Expert Syst. Appl. **37**(10), 6818–6831 (2010)
47. Gotmare, A., Patidar, R., George, N.V.: Nonlinear system identification using a cuckoo search optimized adaptive Hammerstein model. Expert Syst. Appl. **42**, 2538–2546 (2015)
48. Gao, X., Ren, X., Zhu, C., Zhang, C.: Identification and control for Hammerstein systems with hysteresis non-linearity. IET Control Theory Appl. **9**(13), 1935–1947 (2015)
49. Voros, J.: Identification of Hammerstein systems with time-varying piecewise-linear characteristics. IEEE Trans. Circuits Syst. II Express Briefs **52**(12), 865–869 (2005)
50. Greblicki, W.: Non-parametric orthogonal series identification of Hammerstein systems. J. Int. J. Syst. Sci. **20**(12), 2355–2367 (1989)
51. Eskinat, E., Johnson, S.H., Luyben, W.L.: Use of Hammerstein models in identification of nonlinear systems. AIChE J. **37**(2), 255268 (1991)
52. Storn, R., Price, K.: Differential evolution -a simple and efficient adaptive scheme for global optimisation over continuous spaces. Technical Report TR-95–012. ICSI, Berkeley, CA (1995)
53. Kennedy, J., Eberhart, R.: Particle swarm optimization. Proc. 1995 IEEE Int. Conf. Neural Netw. **4**, 1942–1948 (1995)
54. Hachino, T., Deguchi, K,, Takata, S.: Identification of Hammerstein model using radial basis function networks and genetic algorithm. In: Control Conference 2004. Fifth Asian, vol. 1, pp 124–129 (2004)
55. Salomon, R.: Evolutionary algorithms and gradient search: similarities and differences. IEEE Trans. Evol. Comput. **2**(2), 45–55 (1998)
56. Hatanaka, T., Uosaki, K., Koga, M.: Evolutionary computation approach to block oriented nonlinear model identification. In: 5th Asian Control Conference, Malaysia, 1, pp. 90–96 (2004)
57. Wilcoxon, F.: Individual comparisons by ranking methods. Biometrics **1**, 80–83 (1945)
58. Garcia, S., Molina, D., Lozano, M., Herrera, F.: A study on the use of non-parametric tests for analyzing the evolutionary algorithms' behaviour: a case study on the CEC'2005 Special session on real parameter optimization. J Heurist. **15**, 617–644 (2009)

Chapter 6
A States of Matter Search-Based Scheme to Solve the Problem of Power Allocation in Plug-in Electric Cars

Abstract In recent years, researchers have proved that the electrification of the transport sector is vital for reducing both the emission of green-house pollutants and the dependence on oil for transportation. As a result of this, Plug-in Hybrid Electric Vehicles (PHEVs) have received increasing attention during the last decade. A large scale penetration of PHEVs into the marked is expected to take place shortly; however, an unattended increase on the PHEVs needs may yield to several technical problems that could potentially compromise the stability of power systems. As a result of the growing necessity for addressing such issues, topics related to the optimization of PHEVs' charging infrastructures have captured the attention of many researchers. Related to this, several state-of-the-art swarm optimization techniques (such as the well-known Particle Swarm Optimization (PSO) algorithm or the recently proposed Gravitational Search Algorithm (GSA) approach) have been successfully applied in the optimization of the average State of Charge (SoC), which stand as one of the most important performance indicators in the context of PHEVs' intelligent power allocation; however, many of these swarm optimization methods are known to be subject to several critical flaws, such as premature convergence and a lack of balance between the exploration and exploitation of solutions. Such problems are usually related to the evolutionary operators employed by each of such methods on the exploration and exploitation of new solutions. In this chapter, the recently proposed States of Matter Search (SMS) swarm optimization method is proposed for maximizing the average State of Charge of PHEVs within a charging station. In our experiments, several different scenarios consisting of different numbers of PHEVs were considered. In order to test the feasibility of the proposed approach, comparative experiments were performed against other popular PHEVs' State of Charge maximization approaches based on swarm optimization methods. The results obtained on our experimental set up show that the proposed SMS based SoC maximization approach has an outstanding performance in comparison to that of the other compared methods, and as such, proves to be superior for tackling the challenging problem of PHEVs smart charging.

6.1 Introduction

Carbon dioxide (CO_2) is the principal pollutant gas emitted as a result of several human activities related to the combustion of fossil fuels (such as coal, natural gas, and oil). The use of fossil fuel for transportation accounts for about 25% of the CO2 emissions around the world while also considering for over 55% of the world's oil consumption [1]. But, in recent years, the classical view of power systems has been reshaped as a result of technological advances, as well as several economic and environmental considerations. Several kinds of research have proved that significant reductions in both greenhouse gas emissions and the dependence on fossil fuels could be accomplished by the electrification of the transport sector [2]. As a result of this, Plug-in Hybrid Electric Vehicles (or PHEVs) have received never before seen raised attention related to their low emission of pollutants and overall low energy cost. New PHEVs technologies guarantee to improve the overall fuel efficiency by achieving specialized battery systems, which could allow such hybrid vehicles to be charged from traditional power grid systems. In this sense, the currently growing PHEVs tendency promises to shift the current energy demand from fossil fuel to electricity [3].

Statistics provided by the Electric Power Research Institute (EPRI) suggest that by the year 2050, about 62% of the United States vehicle fleet will be comprised of PHEVs [4]. But, an untended increase of PHEV demands may endanger the stability of power systems, and as such, there is an increasing necessity for addressing the technological implications. Moreover, the power demand patterns related to differences in the needs of multiple PHEVs within a charging station have a naturally significant impact on the electric power market. Therefore, it must also be considered [5]. There is a growing need for more efficient algorithms and mechanisms which could allow smart grid technologies to handle complex problems (such as energy management, efficient charging infrastructure, cost reduction, etc.), subject to a wide variety of different objectives and system constraints. To both maximize customer comfort and minimize the inconveniences and changes to the power grid, a specialized control approach is required to manage multiple PHEV battery loads [6] properly. These control mechanisms must consider several real-world constraints, such as variations on the infrastructure and communication among individual vehicles. They must also be suitable to adapt to variations in times between arrivals and departures, as well as the number of PHEVs within the charging station. A critical constraint for the correct charging of PHEVs is related to their battery State-of-Charge. The State-of-Charge (or SoC) is a parameter that measures the amount of electrical energy saved in a vehicle's battery. The performance of PHEVs is highly dependent on the proper management of electric power, which is solely dependent on the battery's SoC.

Latterly, several researchers have concentrated their works on changing the interaction between PHEVs and the electric power grid by proposing a wide variety of intelligent power allocation frameworks. One of the most traditional approaches to solving these power allocation problems requires the use of optimization techniques known as swarm-optimization methods [7]. In [8, 9], respectively, the Particle Swarm

Optimization (PSO) and Gravitational Search Algorithm (GSA) methods have been successfully applied for solving the problem of smart power management among a given number of PHEVs under different SoC conditions. While the PSO and GSA methods are known for providing satisfactory results, such methods tend to produce suboptimal solutions related to a lack of equilibrium between the exploration and exploitation of solutions [10, 11].

In this chapter, the recently developed swarm optimization method known as the State of Matter Search (SMS) is proposed for solving the problem of intelligent power allocation on PHEVs. In the SMS method, individual particles emulate a set of particles that combine in correspondence to a set of evolutionary operators based on the principles of thermal-energy motion [12]. The SMS fuses the use of such operators with a single control strategy based on the states of matter transitions, which modifies the parameter setting of each operation during the evolutionary process. In contrast to other swarm optimization techniques, the computational procedures incorporated in the SMS approach yield a better balance between the exploration and exploitation of solutions, typically leading to better results [12]. To prove its usefulness for solving the proposed optimization problem, the SMS was compared in terms of performance with several other state-of-the-art swarm optimization methods such as PSO and GSA.

6.2 Problem Formulation

The principal motivation after a smart charging scheme is to charge a vehicle whenever it is most useful, such as when the price of electricity and the total power need are low, or when the produced power reaches an excess capacity [13]. For the system to be effective, it should provide most PHEVs to leave the charging station before their expected charging time.

Suppose there is a charging station with a total power capacity P specifically designed to charge Plug-in Hybrid Vehicles (PHEVs). The principal objective is to allot power intelligently for each PHEV that evolves into the charging station. To do so, it is necessary to analyze each PHEV's State-of-Charge (SoC). Each vehicle's SoC changes continually as a result of the charging process, so, it is necessary to maintain track of such changes to allot power effectively. Hence, each vehicle's current SoC describes the main parameter which has to be maximized to guarantee a proper power allocation. To do so, an objective function which leads into account several constraints (such as charging time, current SoC, and price of the energy) is examined for the maximizing the of average State-of-Charge [14].

To illustrate this, let N stand for the number of PHEVs that require to be served on a regular 24-h day. Let k denote a fixed time step for which we aim to maximize the average State-of-Charge $J(k)$. An objective function which models such a maximization case may be defined as follows:

$$J(k) = max \left[\sum_{i=1}^{N} w_i(k) \cdot S_oC_i(k+1) \right] \tag{6.1}$$

where $w_i(k)$ denotes a weighting, term associated to a given PHEV i at time step k, $S_oC_i(k+1)$ stands for the vehicle's State-of-Charge at time step $k + 1$. In the proposed approach, it is assumed that the charging current remains constant for every instant of time. That is:

$$[S_oC_i(k+1) - S_oC_i(k)] \cdot C_i = Q = I_i(k)\Delta t \tag{6.2}$$

It follows that:

$$S_oC_i(k+1) = S_oC_i(k) + \frac{I_i(k)\Delta t}{C_i} \tag{6.3}$$

where the sample time Δt is defined by the charging station operators and $I_i(k)$ is the charging current over Δt. Furthermore, the PHEV's battery unit is modeled as a capacitor circuit, where the battery's power capacity Ci is represented by its respective capacitance value (in Farad).

$$C_i \frac{dV_i}{dt} = I_i \tag{6.4}$$

Therefore, over a small-time interval, the change of voltage could be assumed to be linear, such that:

$$C_i \frac{V_i(k+1) - V_i(k)}{\Delta t} = I_i(k) \tag{6.5}$$

from which it follows that:

$$V_i(k+1) - V_i(k) = \frac{I_i \Delta t}{C_i} \tag{6.6}$$

Also, since our decision variable is the power allocated to each individual PHEV, we replace the current term $I_i(k)$ as follows:

$$I_i(k) = \frac{P_i(k)}{V_i'(k)} \tag{6.7}$$

where $P_i(k)$ represents the amount of electric power assigned to the $i - th$ vehicle on the charging station, while $V_i'(k) = 0.5 \cdot (V_i(k+1) - V_i(k))$ denotes the average voltage between the voltage values at time steps k and $k + 1$. By replacing $I_i(k)$ on Eq. (6.3), it follows that:

$$S_oC_i(k+1) = S_oC_i(k) + \frac{P_i(k)\Delta t}{0.5 \cdot C_i \cdot (V_i(k+1) - V_i(k))} \qquad (6.8)$$

where $(V_i(k+1)$ is obtained by replacing $I_i(k)$ in Eq. (6.6). This yield:

$$V_i(k+1) = \sqrt{2\frac{P_i(k)\Delta t}{C_i} + V_i^2(k)} \qquad (6.9)$$

We may finally represent the objective function $J(k)$ as follows:

$$J(k) = \max_{P_i(k)\,\in\,\mathbb{R}} \left[\sum_{i=1}^{N} w_i(k) \cdot \left(S_oC_i(k) + \frac{P_i(k)\Delta t}{0.5 \cdot C_i \cdot \left(\sqrt{2\frac{P_i(k)\Delta t}{C_i} + V_i^2(k)} - V_i(k) \right)} \right) \right] \qquad (6.10)$$

which is subject to:

$$\sum_i P_i(k) \le P_{utility} \cdot \eta \qquad (6.11)$$

$$0 \le P_i(k) \le P_{i,max} \qquad (6.12)$$

$$0 \le S_oC_i(k) \le S_oC_{i,max} \qquad (6.13)$$

where $P_{utility} = P_{i,max} \cdot N$ denotes the maximum power that can be provided by a charging station with capacity for N PHEVs [15], while $P_{i,max}$ stands for the maximum power that can be absorbed by a specific PHEV. Also, h denotes the efficiency of the charging station (typically of 90%) and it is assumed to be constant. Furthermore, $S_oC_{i,max} = 0.8$ stands for SoC limit related to the i-th PHEV [9] (see Table 6.1). In practice, it is not recommended to fully charge a PHEV's battery due

Table 6.1 Parameter settings for the PHEV's smart power allocation objective function

Parameters	Description	Values
Fixed parameters	PHEV's maximum power absorption Charging station efficiency Total charging time (time step length)	$P_{i,max} = 6.7$ kWh $\eta = 0.9$ $\Delta t = 20\ min$ (1200 s)
Variables	PHEV's State of Charge (SoC) PHEV's battery capacity	$0.2 \le SoC_i \le 0.8$ 16kWh $\le C_i \le 40$ kWh
Constraints	$\sum_i P_i(k) \le P_{utility}(k) \cdot \eta$ $0 \le P_i(k) \le P_{i\,max}(k)$	

to the risk of a possible overload. As such, the value $S_oC_{i,max}$ is set as a security measure in order to prevent damage to the battery [16].

Finally, the charging weight $w_i(k)$ is expressed as a function of three particular parameters, defined as follows:

$$w_i(k) = f(c_i(k), t_i(k), d_i(k)) \tag{6.14}$$

where $c_i(k) = Ci \cdot (1 - S_oC_i(k))$ denotes the proportion of the i-th PHEV's rated battery capacity C_i that remains to be filled at a given time step k. Furthermore, $t_i(k)$ stands for the i-th PHEV's remaining charging time at time step k, and $d_i(k)$ represents the difference between the price of the real-time energy and the price that a specific customer at the i-th PHEV charger is willing to pay at such time step k [17]. In this case, the weighting term $w_i(k)$ gives a degree of preference which is proportional to certain specific attributes of each individual PHEV; i.e., if a given PHEV has both, a lower initial State-of-Charge and less remaining charging time, but the driver is eager to pay a higher price, the system will provide more power to this particular PHEV battery charger:

$$w_i(k) \propto \left[c_i(k) + d_i(k) + \frac{1}{t_i(k)} \right] \tag{6.15}$$

It is worth noting that the terms $c_i(k)$, $d_i(k)$ and $t_i(k)$ are not of the same scale, and as such all terms must be normalized to assign similar relevance to each of them. Furthermore, the charging station operators may also manifest several different interests, which could be influential when assigning an importance factor to each these terms. With that being said, we may express each weighting term as follows:

$$w_i(k) = \alpha_1 c_i(k) + \alpha_2 t_i(k) + \alpha_3 d_i(k) \tag{6.16}$$

where α_1, α_2 and α_3 denotes the importance factors assigned to the terms $c_i(k)$, $d_i(k)$ and $t_i(k)$ respectively, and such that $\sum_j \alpha_j = 1$. In the proposed approach, it is assumed that $\alpha_1 = \alpha_2 = \alpha_3$ at every time step k.

6.3   The States of Matter Search (SMS) Algorithm

Due to its interesting properties, metaheuristic schemes present a good ability to solve complex engineering problems [22–24]. The States of Matter Search (SMS) is a metaheuristic optimization method which emulates the states of matter phenomenon. In this optimization approach, individuals within a population of search agents are represented as molecules which interact with each other by computing a set of unique evolutionary operators based on the physical principles of the thermal-energy motion mechanism [18]. The SMS's evolutionary process is divided in three sequential stages: (1) a gas state in which the molecules experience severe displacements and collisions; (2) a liquid state in which there is a significant reduction of molecular movement; and (3) a solid state in which the force among particles becomes so strong that molecular movement is almost completely inhibited. As the SMS evolutionary process transitions from one stage to another, different movement behaviors are exhibited by the molecules within a given search space. Such behaviors allow the SMS method to preserve a better balance between the exploration and exploitation of solutions, allowing the evolutionary process to find potentially better solutions [19].

6.3.1 States of Matter Transition

As previously stated, in the States of Matter Search (SMS) approach the whole optimization process is divided into three different stages: (1) gas state; (2) liquid state; and (3) solid state. The gas state comprises the first stage of the SMS method. In this stage, molecules experience severe displacements and collisions. The gas state lasts for 50% of the total iterations which comprise the whole SMS's optimization process. The next stage in the SMS optimization process is the liquid state. In this stage, the motion and collisions exhibited by the molecules within the search space are more restricted in comparison to the gas state. The liquid state lasts for 40% of the total iterations of the SMS evolution process. The third and last stage of the SMS optimization method is represented by the solid state. In this stage, forces among particles are much stronger in comparison to the previous SMS stages, which in turn prevents particles from moving freely. The solid state lasts for only the remaining 10% of total SMS's iterations. The overall transition of the SMS optimization process is described in Fig. 6.1.

During each SMS stage, a series of parameters α, β and γ (which are employed on the molecular movement operators described in Sect. 6.3.2) are all modified, allowing SMS to control the way in which the molecules move on each of such stages. Table 6.2 shows the SMS parameters setup corresponding to each particular SMS stage, as given by its own reference [12].

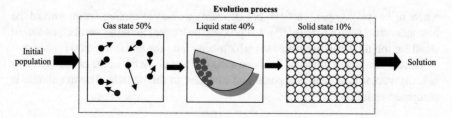

Fig. 6.1 States of Matter Search (SMS) evolution process

Table 6.2 States of Matter Search (SMS) parameters setup

Stage	Duration (%)	α	β	γ	$\mathcal{P}$
Gas	50	0.8	0.8	[0.8, 1.0]	0.9
Liquid	40	0.4	0.2	[0.0, 0.6]	0.2
Solid	10	0.1	0.0	[0.0, 0.1]	0.0

6.3.2   Molecule Movement Operators

In the SMS approach, search agents are treated as molecules whose positions within a multi-dimensional space are changed as the algorithm evolves. In the SMS, the movement of such molecules is analogous to the principles which govern the motion of thermal-energy. The movement of every molecule is determined by considering: (1) the attraction force among the molecules; (2) a series of collision phenomenon; and (3) some random phenomena experienced by the molecules [18].

6.3.2.1   Direction of Movement

The direction vector operator indicates the way in which molecules will change their positions as the SMS's evolution process develops. For each n-dimensional molecule $\mathbf{p}_i$ within the total population of molecules $\mathbf{P}$, an n-dimensional direction vector $\mathbf{d}_i$ is assigned to each particle. During the SMS's initialization process, all direction vectors $(D = \{\mathbf{d}_1, \mathbf{d}_2, \ldots \mathbf{d}_N\})$ are randomly initialized with valves within the range $[-1, 1]$.

As the system develops, the molecules within the search space experience several attraction forces. This attraction force is modeled as a movement towards the best solution found so far by the evolutionary process. Therefore, at each iteration of the evolutionary process, the direction vector update is computed as follows:

$$\mathbf{d}_i^{k+1}1 = \mathbf{d}_i^k \cdot \left(1 - \frac{it}{itern}\right) \cdot 0.5 + \mathbf{a}_i \tag{6.17}$$

where $\mathbf{a}_i = \left(\mathbf{p}^{best} - \mathbf{p}_i\right)/\left\|\mathbf{p}^{best} - \mathbf{p}_i\right\|$ denotes a unitary attraction vector toward the best individual seen so-far $\left(\mathbf{p}^{best}\right)$, k represent the current iteration number, and *itern* stand for the total iterations number which constitute the entire evolution process.

Once a movement direction has been assigned to a given molecule pi its respective velocity vector $\mathbf{v}_i$ is then computed and assigned to the particle. The magnitude is computed as follows:

$$\mathbf{v}_i = v_{init} \cdot \mathbf{d}_i \tag{6.18}$$

where v_{init} denotes a velocity magnitude which is calculated as follows:

$$v_{init} = \alpha \cdot \frac{\sum_{j=1}^{n}\left(b_j^{high} - b_j^{low}\right)}{n} \qquad (6.19)$$

where b_j^{low} and b_j^{high} denote the lower and upper j-th parameter bounds respectively, while n stands for the total number of decision variables (dimensions). Furthermore, $\beta \in [0, 1]$ denotes a scalar factor whose value depends on the current SMS stage (see Table 6.2).

Finally, the position of each molecule for each molecule $\mathbf{p}_i$ is updated as follows:

$$p_{i,j}^{k+1} = p_{i,j}^{k} + v_j \cdot \text{rand}(0, 1) \cdot \left(b_j^{high} - b_j^{low}\right) \cdot \gamma \qquad (6.20)$$

where rand $(0, 1)$ stand for a random number generated within the range of $[0, 1]$, while $\gamma \in [0, 1]$ stands for a scalar factor whose value depends on the current SMS stage (see Table 6.2).

6.3.2.2   Collisions

The collision operator emulates the collision phenomenon that molecules suffer when they interact with each other. The collision operator is only calculated if the distance between two different molecules $\mathbf{p}_i$ and $\mathbf{p}_q$ is smaller than a given collision radius r, calculated as follows:

$$r = \beta \cdot \frac{\sum_{j=1}^{n}\left(b_j^{high} - b_j^{low}\right)}{n} \qquad (6.21)$$

where b_j^{low} and b_j^{high} represent the lower and upper j-th parameter bounds respectively, while n stands for the total number of decision variables (dimensions). Furthermore, $\beta \in [0, 1]$ denotes a scalar factor whose value depends on the current SMS stage (see Table 6.2).

In other words, if $\|\mathbf{p}_i - \mathbf{p}_q\| < r$, it is assumed that molecules $\mathbf{p}_i$ and $\mathbf{p}_q$ have entered into collision; in such a situation, the direction vectors $\mathbf{d}_i$ and $\mathbf{d}_q$ corresponding to each involved particle are exchanged, such that:

$$\mathbf{d}_i = \mathbf{d}_q \text{ and } \mathbf{d}_q = \mathbf{d}_i \qquad (6.22)$$

The collision operator provides SMS the ability to control the diversity of solutions by forcing molecules to change their directions whenever they get close to each other, which prevents them from prematurely overcrowding a given region within the search space.

6.3.2.3 Random Behavior

In order to simulate the random behavior commonly demonstrated by molecules during their transition from one matter state to another, the SMS method integrates an operator which, by following a probabilistic criterion, allows it to randomly change the positions of molecules within a given search space. Under such operation, each molecule within the set of positions $\mathbf{P}^{k+1} = \{\mathbf{p}_1^{k+1}, \mathbf{p}_2^{k+1}, \ldots, \mathbf{p}_N^{k+1}\}$ (which corresponds to the set of positions generated by computing the movement operators described in Sect. 6.3.2.1) is assigned a probability of changing its current position with that of a randomly generated molecule within the feasible search space. This mechanism may be modeled as follows:

$$p_{i,j}^{k+1} = \begin{cases} b_j^{low} + rand(0,1) \cdot \left(b_j^{high} - b_j^{low} \right) with\ probability\ P \\ p_{i,j}^{k+1}\ with\ probability\ (1-P) \end{cases} \tag{6.23}$$

where b_j^{low} and b_j^{high} denote the lower and upper j-th parameter bounds respectively, while rand $(0, 1)$ stand for a random number within the range $[0, 1]$. Furthermore, P represents the probability that a given particle has to change its current position and its value (as mentioned in Sect. 6.3.2) depending on the current SMS stage (see Table 6.2).

6.4 SMS-Based Smart Power Allocation for PHEVs

In this chapter, the States of Matter Search (SMS) algorithm is proposed to solve the problem of smart power allocation for PHEVs. As illustrated in Sect. 6.2, the main objective of a PHEV's smart power allocation algorithm may be described as follow. For a given time step k, find an optimal power allocation configuration $[P_1(k), P_2(k), \ldots, P_N(k)]$ which leads to the average State-of Charge $J(k)$ [as given by Eq. (6.10)] yielding a maximum value.

In the proposed approach, the SMS algorithm starts by generating a set of N molecules $\mathbf{P}(k) = \{\mathbf{p}_1(k), \mathbf{p}_2(k), \ldots, \mathbf{p}_N(k)\}$ within a feasible search space. In the context of a PHEV's smart power allocation problem, the positions occupied by each molecule $\mathbf{p}_j(k) \in \mathbf{P}(k)$ (with k denoting a specific time step) represents a specific power allocation vector, given as follows:

$$\mathbf{p}_j(k) = \left[p_{j,1}(k), p_{j,2}(k), \ldots, p_{j,n}(k) \right] \tag{6.24}$$

where the elements $\mathbf{p}_{j,i}(k)$ represent the total power allocated to the i-th PHEV at a given time step k, and where $j \in \{1, 2, \ldots, N\}$ represents the index of the j-th molecule.

Guided by the SMS's unique evolutionary operators, each molecule $\mathbf{p}_j(k)$ moves around a feasible search space while looking for an optimal power allocation vector configuration. The quality (fitness) of each of molecule $\mathbf{p}_j(k)$ is evaluated by considering the average State of Charge function given by Eq. (6.10), such that:

$$J\left(\mathbf{p}_j(k)\right) = \sum_{i=1}^{N} w_i(k) \cdot \left(S_o C_i(k) + \frac{p_{j,1}(k) \cdot \Delta t}{0.5 \cdot C_i \cdot \left(\sqrt{2 \frac{p_{j,1}(k) \cdot \Delta t}{C_i} + V_i^2(k)} - V_i(k) \right)} \right)$$

(6.25)

where $J\left(\mathbf{p}_j(k)\right)$ represents the average State of Charge computed with regard to the power allocation vector corresponding to the molecule $\mathbf{p}_j(k)$ at a given time step k. Furthermore, since $J\left(\mathbf{p}_j(k)\right)$ is given for a specific time step k (each defined by charging time of length $\Delta t = 20$ min, as given in Table 6.1) within a regular 24-h day, an optimal power allocation vector must be found for each finite time period. Figure 6.2 illustrates an example of several power allocation configurations, assigned to 50 PHEVs during a regular 24-h day.

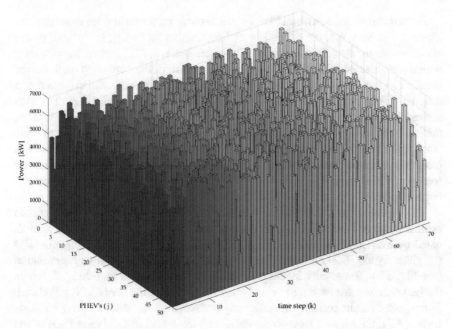

Fig. 6.2 Illustration of different power allocation configurations assigned to 50 PHEVs during a regular 24-h day

6.5 Experimental Results

In order to verify the feasibility and effectiveness of the proposed approach, a series of comparative experiments were performed against the following two state-of-the-art swarm optimization methods:

(1) Particle Swarm Optimization (PSO) method [20]; (2) the Gravitational Search Algorithm (GSA) approach [21]; (3) the Firefly Algorithm (FA) [22]; and (4) the Genetic Algorithms (GA) approach [23]. The parameter settings for each implemented method is described as follows:

(a) PSO: The Standard Particle Swarm Optimization (SPSO-2011) proposed in [20] was implemented. The algorithm's learning factors were set to $c_1 = 2$ and $c_2 = 2$.
(b) GSA: The initial gravitation constant value has been set to $Go = 100$, while the constant parameter alpha has been set to $\alpha = 20$, as given in [21].
(c) SMS: This algorithm was implemented by considering the parameter setup illustrated in Sect. 6.3 (see Table 6.2), as recommended in [12].

For our experiments, several PHEVs' smart charging scenario were simulated. As illustrated in Sect. 6.4, the general procedure consists on finding an optimal power allocation vector at each finite time step k (where $k = 1, 2, …, 72$ when $\Delta t = 20$ min) within a regular 24-h day. With that being said, each of the compared methods are required to run a total of 72 times to complete a single simulation. Furthermore, 5 different PHEVs' power allocation scenarios were also considered. On each of such scenarios, a fixed number of PHEVs at a time (50, 100, 300, 500 and 1000) is taken into account when performing a simulation. Each of such particular scenarios were simulated a total of 50 times each, by considering a population size of 50 individuals (search agents) and a maximum of 100 iterations for each simulation. All calculations were performed on an AMD (R) A6-5400 k CPU 3.60 GHz, 4.0 GB RAM, Microsoft 64bit Windows 7 OS, and MATLAB© R2015A.

The experimental setup aims to compare the proposed approach's performance against those of PSO and GSA. In each of such approaches, performance is evaluated by averaging the values obtained by computing the objective function J(k) (as given by Eq. 6.24) at each time step k of an individual 24-h day simulation (see Fig. 6.3). The results for 50 individual runs are reported in Table 6.2, where the best outcome for each particular PHEVs' smart charging scenario is boldfaced. The reported results consider the following performance indexes: The Average Best-so-far (AB), the Median Best-so-far (MB) and the Standard Deviation (SD) of the best-so-far solution. According to this table, the SMS algorithm performance is superior to those of the other compared methods. Such large difference in performance is intuitively related the SMS's method better trade-off between exploration and exploitation. Furthermore, Fig. 6.4 presents the evolution curves for each particular smart charging scenarios. As illustrated in such figure, GSA has a slower convergence rate in comparison to those of PSO and SMS, with the latter two finding their

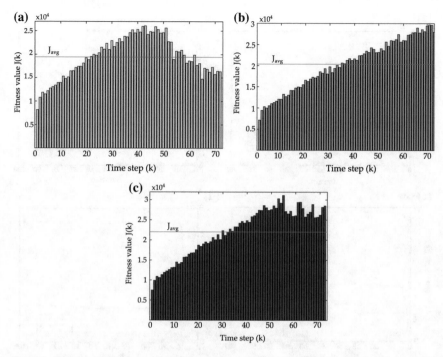

Fig. 6.3 Cost function values $J(k)$ obtained for a simulated PHEVs' smart charging scenario consisting of 50 PHEVs, corresponding to: **a** PSO, **b** GSA and **c** SMS. On each case, the average values J_{avg}, corresponding to the average of all $J(k)$ values obtained during a single simulation, are indicated

best solution in less than 20 iterations on average. However, it is clear that the SMS algorithm surpasses both, PSO and GSA, in terms of solution quality (Table 6.3).

Finally, in Table 6.4, the computational times (in units of seconds) corresponding to each of the compared methods are shown. As evidenced by such table, the computational times corresponding to the SMS method for Finally, in Table 6.3, the computational times (in units of seconds) corresponding to each of the compared methods are shown. As evidenced by such table, the computational times corresponding to the SMS method for

6.6 Conclusions

The swarm optimization method known as States of Matter Search (SMS) was applied to solve the problem of smart power allocation for PHEVs. In the SMS approach, individual molecules that move around a given search space guided by unique evolutionary operators based on the principles of motion of thermal energy. The mechanisms and operators employed by SMS provide a better balance between

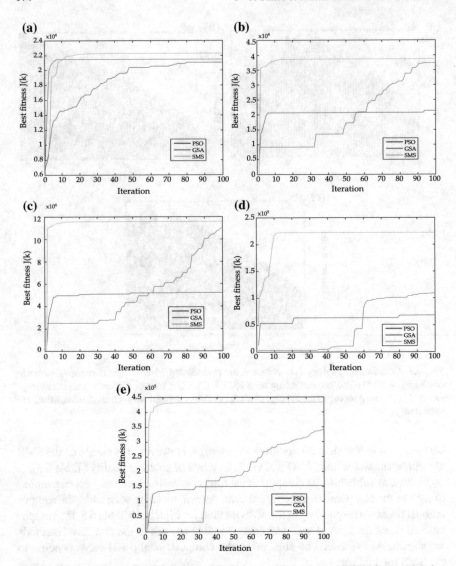

Fig. 6.4 Evolution curves for PSO, GSA and SMS, by considering several PHEVs' smart charging scenarios, consisting of **a** 50 PHEVs, **b** 100 PHEVs, **c** 300 PHEV, **d** 500 PHEVs and **e** 1000 PHEVs

the exploration and exploitation of new solutions, which in turn prevents several issues commonly found in other swarm optimization methods, such as those related to premature convergence.

The performance of the proposed method has been compared to other similar approaches in terms of performance and solution quality. Such comparisons were made by a cost function which takes into account the average state of charge of several hybrid vehicles within a charging station at different time steps of a regular 24-h day.

Table 6.3 Maximization results for several PHEVs' smart charging scenarios, considering n = 50 individual runs and maximum number of iterations itern = 100

Method		Number of PHEVs				
		50	100	300	500	1000
PSO	AB	1.517e+04	3.233e+04	5.117e+04	1.836e+05	1.764e+05
	MB	1.696e+04	3.450e+04	5.433e+04	2.097e+05	2.195e+05
	STD	2.205e+03	4.093e+03	2.339e+03	4.041e+04	7.724e+04
GSA	AB	1.648e+04	3.367e+04	1.156e+05	1.886e+05	3.655e+05
	MB	1.834e+04	3.791e+04	1.1712e+05	2.011e+05	3.760e+05
	STD	3.787e+03	8.139e+03	2.055e+04	2.907e+04	8.045e+04
SMS	AB	**1.864e+04**	**3.939e+04**	**1.214e+05**	**2.067e+05**	**3.892e+05**
	MB	**1.873e+04**	**4.109e+04**	**1.22e+05**	**2.099e+05**	**3.944e+05**
	STD	**9.144e+02**	**1.881e+03**	**4.629e+03**	**6.204e+03**	**1.163e+04**

Table 6.4 Computation time for several different PHEVs' smart charging scenarios. Al reported times are expressed in units of second (sec)

Number of PHEVs	Computational Time (sec)		
	PSO	GSA	SMS
50	45.1584	130.8259	60.8688
100	48.8808	149.8000	69.1776
300	60.6167	237.8983	129.2904
500	70.7352	317.4600	163.6560
1000	99.3896	578.8002	268.8336

Furthermore, several different PHEVs smart charging scenarios were also considered while performing the experimental comparisons. Experimental results show that, compared to the other compared methods, the proposed SMS-based PHEVs' smart power allocation approach yields significantly better results in terms of both performance and solution quality, which further proves the proficiency of the proposed approach for solving the complex problem of smart power allocation for PHEVs.

References

1. Energy Technology Perspectives: Scenarios & Strategies to 2050. Available online: https://www.iea.org/textbase/npsum/etp.pdf. Accessed on 6 Nov 2016
2. Markel, T., Smith, K., Pesaran, A.A.: Improving petroleum displacement potential of PHEVs using enhanced charging scenarios. In: Proceedings of the EVS-24 International Battery, Hybrid and Fuel Cell Electric Vehicle Symposium, Stavanger, Norway, 13–16 May 2009
3. Fazelpour, F., Vafaeipour, M., Rahbari, O., Rosen, M.A.: Intelligent optimization to integrate a plug-in hybrid electric vehicle smart parking lot with renewable energy resources and enhance grid characteristics. Energy Convers. Manag. **77**, 250–261 (2014)

4. Soares, J., Sousa, T., Morais, H., Vale, Z., Canizes, B., Silva, A.: Application-specific modified particle swarm optimization for energy resource scheduling considering vehicle-to-grid. Appl. Soft Comput. **13**, 4264–4280 (2013)
5. Su, W., Chow, M.Y.: Investigating a large-scale PHEV/PEV parking deck in a smart grid environment. In: Proceedings of the 2011 North American Power Symposium, Boston, MA, USA, 4–6 August 2011, pp. 1–6
6. Su, W., Chow, M.Y.: Computational intelligence-based energy management for a large-scale PHEV/PEV enabled municipal parking deck. Appl. Energy **96**, 171–182 (2012)
7. Yang, X.S.: Nature-Inspired Optimization Algorithms. Elsevier, London (2014)
8. Bhattacharyya, S., Dutta, P.: Handbook of Research on Swarm Intelligence in Engineering. IGI Global, Hersey, PA (2015)
9. Rahman, I., Vasant, P.M., Singh, B.S.M., Abdullah-Al-Wadud, M.: On the performance of accelerated particle swarm optimization for charging plug-in hybrid electric vehicles. Alex. Eng. J. **55**, 419–426 (2016)
10. Qin, J., Yin, Y., Ban, X.: A hybrid of particle swarm optimization and local search for multimodal functions. Lect. Notes Comput. Sci. **6145**, 589–596 (2010)
11. Darzi, S., Tiong, S.K., Islam, M.T., Soleymanpour, H.R., Kibria, S., Barrettand, M., Arnott, R., Capon, J., Dahrouj, H., Yu, W., et al.: An experience oriented-convergence improved gravitational search algorithm for minimum variance distortionless response beamforming optimum. PLoS ONE **11**, e0156749 (2016)
12. Cuevas, E., Echavarría, A., Ramírez-Ortegón, M.A.: An optimization algorithm inspired by the states of matter that improves the balance between exploration and exploitation. Appl. Intell. **40**, 256–272 (2014)
13. Chang, W.Y.: The state of charge estimating methods for battery: a review. ISRN Appl. Math. **2013**, 953792 (2013)
14. Rahman, I., Vasant, P.M., Singh, B.S.M., Abdullah-Al-Wadud, M.: Swarm intelligence-based smart energy allocation strategy for charging stations of plug-in hybrid electric vehicles. Math. Probl. Eng. **2015**, 620425 (2015)
15. Su, W., Chow, M.Y.: Performance evaluation of a PHEV parking station using particle swarm optimization. In: Proceedings of the 2011 IEEE Power and Energy Society General Meeting, Detroit, MI, USA, 24–29 July 2011, pp. 1–6
16. Young, K., Wang, C., Wang, L.Y., Strunz, K.: Electric Vehicle Battery Technologies. Springer, New York, NY (2013)
17. Samadi, P., Mohsenian-Rad, A.H., Schober, R., Wong, V.W.S., Jatskevich, J.: Optimal real-time pricing algorithm based on utility maximization for smart grid. In: Proceedings of the 2010 First IEEE International Conference on Smart Grid Communications, Gaithersburg, MD, USA, 4–6 October 2010, pp. 415–420
18. Cengel, Y.A., Boles, M.A.: Thermodynamics: An Engineering Approach, 5th edn. McGraw-Hill, New York, NY (2004)
19. Cuevas, E., Echavarría, A., Zaldívar, D., Pérez-Cisneros, M.: A novel evolutionary algorithm inspired by the states of matter for template matching. Expert Syst. Appl. **40**, 6359–6373 (2013)
20. Zambrano-Bigiarini, M., Clerc, M., Rojas, R.: Standard particle swarm optimisation 2011 at CEC-2013: A baseline for future PSO improvements. In: Proceedings of the 2013 IEEE Congress on Evolutionary Computation, Cancun, Mexico, 20–23 June 2013, pp. 2337–2344
21. Rashedi, E., Nezamabadi-Pour, H., Saryazdi, S.G.S.A.: A Gravitational Search Algorithm. Inf. Sci. **179**, 2232–2248 (2009)
22. Cuevas, E.: Block-matching algorithm based on harmony search optimization for motion estimation. Appl. Intell. **39**(1), 165–183 (2013)
23. Díaz-Cortés, M.-A., Ortega-Sánchez, N., Hinojosa, S., Cuevas, E., Rojas, R., Demin, A.: A multi-level thresholding method for breast thermograms analysis using Dragonfly algorithm. Infrared Phys. Technol. **93**, 346–361 (2018)
24. Díaz, P., Pérez-Cisneros, M., Cuevas, E., Hinojosa, S., Zaldivar, D.: An improved crow search algorithm applied to energy problems. Energies **11**(3), 571 (2018)

Chapter 7
Locus Search Method for Power Loss Reduction on Distribution Networks

Abstract Power losses are presents in the distribution of energy from sources to points of consumption. This power loss is commonly caused by the lack of reactive power. The installation of capacitor banks is an alternative to provided reactive power compensation to Radial Distribution Networks (RDNs). A suitable allocation of capacitor banks brings several advantages to the RDNs, comprising a power loss reduction and enhancement in the voltage profile in the system. These improvements lead to significant energy savings and cost reductions. From an optimization view, this problem is known in literature as Optimal Capacitor Allocation (OCA) and is conceptualized significantly complex by its discontinuity, non-linearity and high multi-modality. In this work, the OCA problem is addressed by the swarm optimization algorithm Locust Search (LS). To measure the performance of the proposed method, it has been tested over diverse and IEEE's radial distribution test system and compared against other related methodologies currently reported in literature to solve the OCA.

7.1 Introduction

The voltage imbalance phenomenon in RDNs is a problem associated with power generation systems that the science community has widely studied. This problem is characterized by a gradual decay on power, induced by the inability of distribution feeders to satisfy the ever-increasing power needs of domestic, commercial, and industrial loads [1]. These voltage instabilities are generally caused by abrupt variations in the systems operating conditions, which increase the demand for reactive power [2]. In order to provide reactive power compensation to RDNs, shunt capacitor banks are commonly installed within these distribution systems [2, 3]. A usual solution to provide reactive power compensation to RDNs is the installation of shun capacitors within these distribution systems [2, 3]. The quantity of reactive power compensation added to the network by the installation of shunt capacitors is directly associated with the number of banks, its sizes, type, and placement within the RDNs, which makes determining those factors as a fundamental task to maximize

© Springer Nature Switzerland AG 2021
E. Cuevas et al., *Metaheuristic Computation: A Performance Perspective*,
Intelligent Systems Reference Library 195,
https://doi.org/10.1007/978-3-030-58100-8_7

the benefits. Proper allocation of capacitor banks gives the network several benefits, including an improvement feeder´s voltage and power loss reduction, which leads to energy savings and cost reductions [4]. The OCA is a combinatorial optimization problem well-studied in which its complexity is correlated to the structure and size of the distribution system. Considering a distribution network constructed by n buses and a set of q available capacitor sizes, there exists a collection of $(q + 1)^n$ potential solutions (combinations) that could be physically installed to solve the OCA problem. Computationally, the evaluation of this collection of solutions (even for a medium-sized distribution system) is too costly to be evaluated in a reasonable time. Therefore, this problem is usually solved by using optimization techniques [4, 5].

Previous works tried to solve the OCA problem are identified for their extensive use of analytical methods to model the location and size of capacitor banks. Nevertheless, these methods are also known due to their frequent use of unrealistic assumptions about the distribution systems operating conditions [6–10]. Furthermore, given that these techniques generally consider the size and location as continuous variables, it is common that the results do not match with the physically available standard capacitor sizes and location nodes, respectively [3]. The use of numerical programming methods has been motivated by the difficulties in the analytical methods to solve the OCA. Some of the most prosperous numerical programming methods applied to face this problem include approaches based on mixed-integer programming [11], dynamic programming [12], and local variations [13], just to mention a few. Although these methods have shown to be capable of solving the OCA problem, they keep with several limitations, such as the difficulty of leaving local minima and a notoriously higher computational cost in comparison to analytical methods [14, 15]. From an optimization perspective, the OCA problem is considered complex due to its high multi-modality, non-linearity, and discontinuity [4]. These characteristics make problematic to solve it by using standard optimization techniques [5].

In recent years, the metaheuristic methods have increased the interest as an alternative in solving engineering problems. The metaheuristic approaches have certain advantages over traditional techniques because there are distinguished for their versatility to generically solve different problems independently of their characteristics, while traditional methods need many implementation requirements (such as continuity, convexity, and differentiability. Under such conditions, diverse metaheuristics approaches have been proposed to deal with the OCA formulation. Some examples include Genetic Algorithms (GA) [2], Particle Swarm Optimization (PSO) [16], Bacterial Foraging Optimization Algorithm (BFOA) [17], Gravitational Search Algorithm (GSA) [1], Crow Search Algorithm [3, 18] among others. Despite their acceptable results, these kinds of search strategies exhibit significant problems when they try to solve high multi-modal optimization problems. However, many of these techniques present complications for balancing the diversification and intensification of the produced solutions. As a consequence, solutions tend to either separate uncontrollably (thus preventing them from further refining potentially good solutions) or to converge toward local optima rapidly [14, 15, 19].

Every metaheuristic method has certain particularities to meet the requirements of particular problems.

Each metaheuristic approach possesses unique characteristics to appropriately fulfill the requirements of particular problems. Nevertheless, no single metaheuristic algorithm can solve all problems in a competitive manner. In order to know the best metaheuristic method to solve the OCA problem, a correct evaluation of its advantages and limitations must be conducted. Under such circumstances, the Locust Search (LS) approach is a recent optimization technique which emulates the characteristic foraging behavior manifested by the individuals of locust swarms [20]. Different to other nature-inspired metaheuristics schemes, the LS algorithm explicitly avoids the concentration of individuals around the best during the search process. This important characteristic allows to amend the for several common flaws found in other similar techniques when they face multi-modal optimization problems such as the lack of balance between exploration and exploitation, premature convergence and divergence.

Motivated by these interesting properties, in this chapter, the LS method is proposed to solve the high multi-modal OCA problem in distribution networks. The proposed approach has been tested by considering three specific radial distribution test systems, specifically the IEEE's 10-bus, 33-bus and 69-bus RDN models [3, 21, 22]. Furthermore, in order to prove the feasibility of the proposed LS-based OCA method, our experimental results have also been compared against those produced by other techniques currently reported on the literature. Experimental results suggest that the proposed LS-based method is able to competitively solve the OCA problem in terms of accuracy and robustness.

The remainder of this chapter is organized as follows: in Sect. 7.2 we discuss the particularities related to the OCA problem; in Sect. 7.3 specific details about the LS swarm optimization algorithm are highlighted; in Sect. 7.4, we illustrate our proposed LS-based OCA methodology; in Sect. 7.5 we show our experimental setup and results. Finally, in Sect. 7.6, conclusions are drawn.

7.2 Capacitor Allocation Problem Formulation

OCA represents a combinatorial optimization problem in which is determined the number, location, type and size of shunt capacitors contained in RDNs. In order to solve this problem, many different objective functions have been proposed. In general, the main objective is to minimize the network's total operation costs by providing reactive compensation to already installed distributions feeders. The idea is to reduce the total network losses while is also maintained the buses voltages within certain permissible limits. In this work, two objective functions are considered: the first is a classical function introduced in [23], which is given as follows:

$$C = K_p \times P_{Loss} + \sum_{j=1}^{n} k_j^c \times Q_j^c, \qquad (7.1)$$

where C denotes the total cost (\$) for a radial distribution network comprised by n buses. Furthermore, K_p represents the equivalent cost per unit of power loss (\$/kW), while P_{Loss} stands for the total power loss (kW) of the distribution system. Finally, Q_j^c denotes the size of the shunt capacitor (kVAR) installed at bus 'j', while k_j^c indicates its corresponding cost per unit of reactive power (\$/kVAR). It is worth mentioning that in the cost function presented in Eq. 7.1 the operating, installation and maintenance costs are neglected.

The second objective function considered in our study was recently proposed in [24] and, unlike the model presented by Eq. 7.1, it also considers installation costs for the calculation of the networks total cost. This objective function is given as follows:

$$C = K_p \times t \times P_{Loss} + K_{ic} \times N_q + \sum_{j=1}^{n} k_j^c \times Q_j^c, \qquad (7.2)$$

where K_p is the cost for unit of power loss per hour (\$/kW), t is the time period in hours, K_{ic} is the installation cost (\$), N_q is the number of compensation capacitors to install, K_c correspond to the cost per unit of reactive power (\$/kVAR). For this model, it is also assumed that the maximum reactive power compensation that can be provided to a given electrical distribution system is restricted to be at the most equal to the total reactive power provided by the uncompensated distribution network [24], this is:

$$\sum_{j=1}^{n_c} Q_j^c \le \sum_{i=1}^{n} Q_i^c, \qquad (7.3)$$

where n_c denote the number of compensated buses, while Q_i^c stand for the reactive power value at a given bus 'i' [25].

7.2.1   Power Loss Calculation

In order to calculate the total power loss P_{Loss} (as described in Eq. 7.1), we apply the simplified calculation method developed in [23]. In this approach, a set of simplified line flow equations are proposed as an alternative for calculating the exact line flow, which is known to require a significant amount of iterations. Assuming a single line diagram of three-phase balanced distribution system (see Fig. 7.1), the set of equations is determined as follows:

$$P_i = P_{i+1} + P_{Li} + R_{i,i+1} \times \left(\frac{P_i^2 + Q_i^2}{|V_i|^2} \right)$$
$$\cong P_{i+1} + P_{Li} + R_{i,i+1} \times \left((P_{i+1} + P_{Li})^2 + (Q_{i+1} + Q_{Li})^2 \right) \qquad (7.4)$$

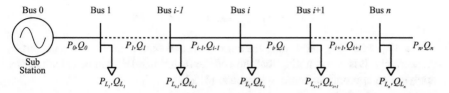

Fig. 7.1 Schematic diagram of a typical radial distribution network comprised by n nodes

$$Q_i = Q_{i+1} + Q_{Li} + X_{i,i+1} \times \left(\frac{P_i^2 + Q_i^2}{|V_i|^2} \right)$$
$$\cong Q_{i+1} + Q_{Li} + X_{i,i+1} \times \left((P_{i+1} + P_{Li})^2 + (Q_{i+1} + Q_{Li})^2 \right) \qquad (7.5)$$

$$|V_{i+1}|^2 = |V_i|^2 - 2 \left(R_{i,i+1} \times P_i + X_{i,i+1} \times Q_i \right) + \left(\frac{P_i^2 + Q_i^2}{|V_i|^2} \right), \qquad (7.6)$$

where P_i denotes the real line power which flows out from bus 'i', while P_{Li} indicates the real load power at bus 'i', and where $R_{i,i+1}$ stands for the line resistance between buses 'i' and '$i + 1$'. Similarly, Q_i represents the reactive line power flowing from bus 'i', while Q_{Li} stands for its respective reactive load power, and where $X_{i,i+1}$ is the line reactance between buses 'i' and '$i + 1$'. Furthermore, $|V_i|$ indicates the voltage magnitude at bus 'i'.

By considering the above formulations, the power loss corresponding to the line connecting buses 'i' and '$i + 1$' is calculated as follows:

$$P_{Loss}^{i,i+1} = R_{i,i+1} \times \left(\frac{P_i^2 + Q_i^2}{|V_i|^2} \right). \qquad (7.7)$$

Finally, the total power loss for the n buses radial distribution network is obtained by adding up the power losses of each line; this is:

$$P_{Loss} = \sum_{i=0}^{n} P_{Loss}^{i,i+1}. \qquad (7.8)$$

7.2.2 Voltage Constrains

For safety reasons, the magnitude of voltage at each bus should be kept within certain permissible limits [4, 7, 12]. This implies that the cost function proposed to solve the OCA problem must also be subjected to these voltage constraints. Therefore, the voltage constraints for each of the buses in the RDN are given as follows:

$$V_{\min} \leq |V_i| \leq V_{\max}, i = 1, 2, \ldots, n, \tag{7.9}$$

where $V_{\min}$ and $V_{\max}$ represents the minimum and maximum bus voltage limits, respectively. It is worth noting that the voltage magnitude $|V_i|$ can be calculated by applying an appropriate iterative approach [3, 23].

7.3 The Locust Search Algorithm

Metaheuristic methods are important tools to solve complex engineering problems [26–28]. The Locust Search algorithm is a metaheuristic optimization technique inspired by the interesting foraging behaviors commonly observed in swarms of locusts [20]. In LS, the solution space is metaphorically represented as a plantation, in which individuals within a swarm of locusts are subjected to constant interactions. In LS, each individual is assigned to a food quality index which is related to the solution that it represents [20]. Another distinctive characteristic about the LS method is the fact that the movement of individuals within the swarm is guided by a set of operators inspired by two distinctive foraging biological behavior phases commonly observed in swarms of locust, namely: (1) Solitary phase, and (2) Social phase.

7.3.1 LS Solitary Phase

Under the solitary phase, each individual within the swarm of locusts is assumed to be displaced as a result of a certain social force exerted over them. The magnitude and orientation of the social force is related to positional relationships between each individual and all other members within the swarm. Also, it is established that the net effect caused by this social force could be manifested as either an attraction toward distant individuals, or a repulsion between nearer individuals [20, 29]. In the LS approach, the conceptualization of social forces was taken as inspiration to develop a movement operator specifically devised to explicitly avoid the concentration of individual toward the best solutions found so far by the search process, which in turn allows a more diverse exploration of the available solution space [20, 29].

In order to illustrate the LS's social phase movement operator, let $\mathbf{L}^k = \{\mathbf{l}_1^k, \mathbf{l}_2^k, \ldots, \mathbf{l}_N^k\}$ denote a population (set of solutions) comprised by N locust positions (where $k = 1, 2, \ldots, itern$ denotes the current iteration of the LS search process and where $itern$ denotes the maximum number of iterations). At each iteration k, the solitary movement operator produces a new candidate solution $\mathbf{p}_i^k$ by perturbing the current locust position $\mathbf{l}_i^k$ as follows:

$$\mathbf{p}_i^k = \mathbf{l}_i^k + \Delta \mathbf{l}_i^k, \tag{7.10}$$

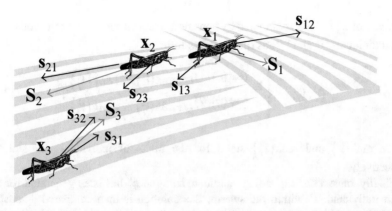

Fig. 7.2 LS solitary phase. Under this behavioral model, the movement of each locust individual 'i' is computed with regard to the total social force that all other members of the population exert toward it

where $\Delta \mathbf{l}_i^k$ represents a position change vector which components equals those of the total social force $\mathbf{S}_i^k$ experimented by the i-th locust individual (hence, $\Delta \mathbf{l}_i^k = \mathbf{S}_i^k$). This social force is roughly represented as:

$$\mathbf{S}_i^k = \sum_{j \neq i} \mathbf{s}_{ij}^k, \tag{7.11}$$

where $\mathbf{s}_{ij}^k$ represents the individual social force exerted toward some individual 'i' by some other individual 'j' (see Fig. 7.2) as given as follows:

$$\mathbf{s}_{ij}^k = \rho\left(\mathbf{l}_i^k, \mathbf{l}_j^k\right) \cdot s\left(r_{ij}\right) \cdot \mathbf{d}_{ij} + \text{rand}(1, -1), \tag{7.12}$$

where $r_{ij} = \mathbf{l}_i^k - \mathbf{l}_j^k$ stands for the Euclidian distance between the pair of locusts 'i' and 'j', while $\mathbf{d}_{ij} = \left(\mathbf{l}_j^k - \mathbf{l}_i^k\right)/r_{ij}$ represent a unit vector which points from $\mathbf{l}_i^k$ to $\mathbf{l}_j^k$, and where $\text{rand}(1, -1)$ stands for a random number drawn from the uniformly distributed interval $[1, -1]$. Furthermore, $s\left(r_{ij}\right)$ represents the so-called social relation between individuals 'i' and 'j', as given by the following equation:

$$s\left(r_{ij}\right) = F \cdot e^{-r_{ij}/L} - e^{-r_{ij}}, \tag{7.13}$$

where F and L denote, an attraction magnitude and an attractive length scale, respectively [20]. On the other hand, $\rho\left(\mathbf{l}_i^k, \mathbf{l}_j^k\right)$ stand for what is referred as the dominance value among individuals 'i' and 'j'. In the LS approach, each solution within $\mathbf{L}^k$ ($\{\mathbf{l}_1^k, \mathbf{l}_2^k, \ldots, \mathbf{l}_N^k\}$) is assigned with a rank from 0 (zero) to $N - 1$ depending on their current fitness value; under this approach, it is said that best individual within the swarm receives a rank of 0, while the worst individual is ranked as '$N-1$'. Therefore, the value of $\rho\left(\mathbf{l}_i^k, \mathbf{l}_j^k\right)$ is computed by considering the current rank of the considered pair individuals 'i' and 'j', as illustrated as follows:

$$\rho\left(\mathbf{l}_i^k, \mathbf{l}_j^k\right) = \begin{cases} e^{-\left(\mathrm{rank}\left(\mathbf{l}_i^k\right)/N\right)} \, if \, rank\left(\mathbf{l}_i^k\right) < rank\left(\mathbf{l}_j^k\right) \\ e^{-\left(\mathrm{rank}\left(\mathbf{l}_j^k\right)/N\right)} \, if \, rank\left(\mathbf{l}_i^k\right) > rank\left(\mathbf{l}_j^k\right) \end{cases}, \qquad (7.14)$$

where $\mathrm{rank}\left(\mathbf{l}_i^k\right)$ and $\mathrm{rank}\left(\mathbf{l}_j^k\right)$ stand for the ranks of individuals 'i' and 'j', respectively.

Finally, once a corresponding candidate solution $\mathbf{p}_i^k$ has been generated for any given individual 'i' within the swarm, this solution is then compared against its originating positions $\mathbf{l}_i^k$ in terms of solution quality, and then the following position update rule is applied in order to assign the positions of each individual 'i' for the following iteration:

$$\mathbf{l}_i^{k+1} = \begin{cases} \mathbf{p}_i^k \;\; if \; f\left(\mathbf{p}_i^k\right) > f\left(\mathbf{l}_i^k\right) \\ \mathbf{l}_i^k \;\; otherwise \end{cases}, \qquad (7.15)$$

where $f\left(\mathbf{p}_i^k\right)$ and $f\left(\mathbf{l}_i^k\right)$ denotes the fitness evaluation function corresponding to $\mathbf{p}_i^k$ and $\mathbf{l}_i^k$, respectively. It is important to note that the previous is illustrated by considering a maximization optimization problem.

7.3.2 LS Social Phase

Different to the solitary phase, the social phase operator represents a selective operation to refine a specific subset of individuals $\mathbf{B} = \{\mathbf{b}_1, \mathbf{b}_2, \ldots, \mathbf{b}_q\}$ within the swarm's population in order to improve their solution quality [20, 29]. Said subset $\mathbf{B}$ is formed by the q best individuals within the set of solutions $\mathbf{L}^{k+1} = \{\mathbf{l}_1^{k+1}, \mathbf{l}_2^{k+1}, \ldots, \mathbf{l}_N^{k+1}\}$, which is represented by the positions adopted by each locust in the swarm after applying the solitary phase movement operator (see Sect. 7.3.1).

To apply the social phase operator, a corresponding subspace C_j around each individual within the subset $\mathbf{B} \in \mathbf{L}^{k+1}$ is created. The limits of each of subspace is defined by a certain parameter r, which is given as follows:

$$r = \frac{\sum_{d=1}^{n}\left(b_d^{high} - b_d^{low}\right)}{n} \cdot \beta, \qquad (7.16)$$

where b_d^{low} and b_d^{high} denote the lower and upper bounds corresponding to the d-th dimension, respectively, while n stand for the total number of decision variables (dimensions). Finally, $\beta \in [0, 1]$ represents a scalar factor. Therefore, for each solution $\mathbf{l}_j^{k+1} = \left[l_{j,1}^{k+1}, l_{j,2}^{k+1}, \ldots, l_{j,n}^{k+1} \right]$ (with $\mathbf{l}_j^{k+1} \in \mathbf{B}$), the limits for each subspace C_j are given as:

$$C_{j,d}^{low} = l_{j,d}^{k+1} - r$$
$$C_{j,d}^{high} = l_{j,d}^{k+1} + r, \qquad (7.17)$$

where $C_{j,d}^{low}$ and $C_{j,d}^{high}$ represent the upper and lower bounds of each subspace C_j at the d-th dimension, respectively.

Further, for each solution $\mathbf{l}_j^{k+1} \in \mathbf{B}$, a set of h new candidate solutions $\mathbf{M}^j = \left\{ \mathbf{m}_1^j, \mathbf{m}_2^j, \ldots, \mathbf{m}_h^j \right\}$ is generated within the bounds of their corresponding subspace C_j (see Fig. 7.3), and then their solution quality is evaluated; finally, if the fitness value of any solutions $\mathbf{m}_i^j \in \mathbf{M}^j$ is better than that of their originating solution $\mathbf{l}_j^{k+1}$, then $\mathbf{l}_j^{k+1}$ is replaced by $\mathbf{m}_i^j$; otherwise, no changes are made to $\mathbf{l}_j^{k+1}$. The previous solution update rule may be summarized by the following expression:

$$\mathbf{l}_i^{k+1} = \begin{cases} \mathbf{m}_i^j & if \left(f\left(\mathbf{m}_i^j\right) = \max_n \left\{ f\left(\mathbf{m}_n^j\right) \right\} \right) and \left(f\left(\mathbf{m}_i^j\right) > f\left(\mathbf{l}_i^{k+1}\right) \right) \\ \mathbf{l}_i^{k+1} & otherwise \end{cases}, \quad (7.18)$$

where the $\mathbf{l}_j^{k+1}$ belongs to the subset of best solutions $\mathbf{B}$ ($\mathbf{l}_j^{k+1} \in \mathbf{B}$) [20, 29].

(a)

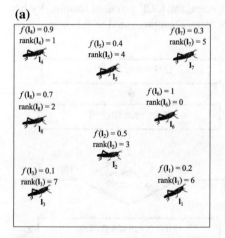

(b)

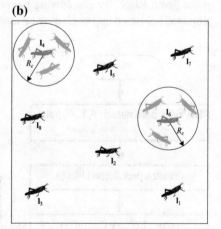

Fig. 7.3 Illustration of the LS social phase. **a** Initial configuration and rankings for the members of a swarm comprised by N = 8 locusts, and **b** Implementation of the social phase operator by considering q = 2 and h = 3

7.4   Optimal Capacitor Allocation Based on LS-Algorithm

In this chapter, the LS algorithm (as described in Sect. 7.3) is proposed for solving the OCP problem. As illustrated in Sect. 7.2, the OCP problem can be modeled as a combinatorial optimization problem in which the objective is to minimize the total operation cost on RDNs by placing shunt capacitors of specific sizes on the distribution feeders of the electrical system.

While the LS algorithm was initially proposed to solve continuous global optimization problem, we have modified it so that it can properly handle the OCP problem. The modification applied to the LS algorithm consists on rounding the real-coded elements provided by each candidate solution to their nearest integer value (see Fig. 7.4). While this modification is conceptually simple, its practicity contributes to properly handle the OCP problem while also avoiding extensive modifications to the original search method [30]. With that being said, our modified LS algorithm initializes by generating a population of N solutions, each represented by a corresponding candidate solution of the form $\mathbf{Q} = [q_1, q_2, \ldots q_n]$. From an OCP perspective, each component q_j from $\mathbf{Q}$ represents the index number of a specific commercial shunt capacitor that is to be placed on the j-th bus of a given n-buses RDN. In this sense, it is considered that each indexed capacitor has a reactance value $Q_j = Q(q_j)$ that match with that of a commercial shunt capacitor (see Table 7.1) [31]. Also, it worth nothing that in our proposed approach, all buses of a given RDN are considered as potential candidates for the allocation of shunt capacitors. Finally, and as a special consideration, if a capacitor index value of '0' (cero) is assigned to any component q_j of $\mathbf{Q}$, then it is assumed that no capacitor is placed at all on the specified bus 'j' of the given RDN. By considering the previous, the OCP problem handled by our proposed LS-based approach may be more formally represented as follows:

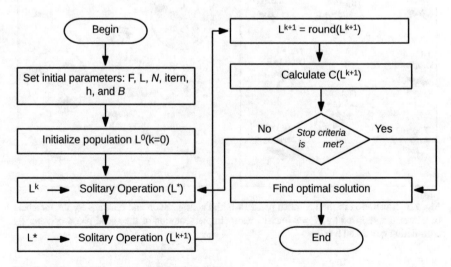

Fig. 7.4 Flowchart for the modified LS-algorithm

Table 7.1 Commercial capacitor sizes (kVAR) and their corresponding costs per unit of reactive power ($/kVAR)

Capacitor index (No.)	Capacitor size (kVAR)	Cost per unit of reactive power ($/kVAR)
q_j	$Q_j = Q(q_j)$	$k(Q_j)$
0	0	0
1	150	0.500
2	350	0.350
3	450	0.253
4	600	0.220
5	750	0.276
6	900	0.183
7	1050	0.228
8	1200	0.170
19	1350	0.207
10	1500	0.201
11	1650	0.193
12	1800	0.187
13	1950	0.211
14	2100	0.176
15	2250	0.197
16	2400	0.170
17	2550	0.189
18	2700	0.187
19	2850	0.183
20	3000	0.180
21	3150	0.195
22	3300	0.174
23	3450	0.188
24	3600	0.170
25	3750	0.183
26	3900	0.182
27	4050	0.179

In the case of the objective function presented by Eq. 7.1, the cost minimization problem may be expressed as:

Minimize

$$C(\mathbf{Q}) = K_{\mathrm{p}} \times P_{\mathrm{Loss}}(\mathbf{Q}) + \sum_{i=1}^{n} Q_j \times k(Q_j),$$

Subject to:

$$Q_j \in \{0, 150, 360, \ldots, 3900, 4050\}, \tag{7.19}$$

while for the case of the cost function presented in Eq. 7.2 this will be:

Minimize

$$C(\mathbf{Q}) = K_p \times t \times P_{\text{Loss}}(\mathbf{Q}) + K_{ic} \times N_q + \sum_{j=1}^{n} Q_j \times k(Q_j),$$

Subject to:

$$Q_j \in \{0, 150, 360, \ldots, 3900, 4050\},$$

$$\sum_{j=1}^{n_c} Q_j^c \leq \sum_{i=1}^{n} Q_i^c, \tag{7.20}$$

where $k(Q_j)$ indicates its corresponding cost (see Table 7.1), and where:

$$P_{\text{Loss}}(\mathbf{Q}) = \sum_{i=0}^{n} P_{\text{Loss}}(Q_i), \tag{7.21}$$

where $P_{\text{Loss}}(Q_i) = P_{\text{Loss}}^{i,i+1}$ stand for the power loss corresponding to the line connecting buses 'i' and '$i + 1$', as given by Eq. 7.5.

7.5 Experimental Results

In the previous section, an OCP method based on the swarm optimization algorithm known as LS has been proposed. In order to prove the effectiveness of the proposed LS-based OCP scheme, it has been tested by considering three particular RDNs models commonly reported on the literature as reference, namely the IEEE's 10-Bus, 33-Bus and 69-Bus RDNs [3, 21, 32]. Also, the results produced by our proposed OCP method for each of the considered RDN models are further compared against those produced by other techniques currently reported on the literature. All test RDNs has been tested with regard to the two objective functions presented in Sect. 7.2 (the specific parameters applied for each function are shown in Table 7.2). For each case of study, the LS algorithm has been executed a total of 30 times, each by considering $K = 1000$ iterations and a population size of $N = 20$ individual (locusts). All experiments were performed on MatLAB® R2017a, running on a computer with Pentium Dual-Core 2.5 GHz processor, and 8-GB of RAM.

Table 7.2 Parameters for objective functions

Parameters	Case 1	Case 2
K_p	$168	$0.06
k_j^c	See Table 7.1	–
t	–	8760 h
K_{ic}	–	$1000
K_c	–	$3.000

Table 7.3 Statistical results for the proposed LS-Based OCP approach

Case	RDN	Min	Max	Avg	Std
1	10-Bus	115,490.8	115,715.1	115,522.5	0.0006
	33-Bus	23,636.1	23,706.9	23,609.9	0.0050
	69-Bus	24,825.2	25,033.4	24,922.9	0.0065
2	10-Bus	370,929.7	376,629.6	374,394.3	0.1050
	33-Bus	81,279.3	82,670.5	81,492.7	0.4300
	69-Bus	83,708.2	83,897.3	83,720.1	0.0030

7.5.1 Statistical Results of Test Cases

In order to validate the robustness of the proposed method, the statistical results corresponding to 30 individual runs are presented in Table 7.3. In this table, the minimum (min), maximum (max), average (avg) and standard deviation (std) of the best results obtained in the minimization of the objective functions presented in Eq. 7.19 (case 1) and 20 (case 2) are shown.

7.5.2 Comparative Results for IEEE's 10-Bus Distribution System

Our first set of experiments involves the IEEE's 10-Bus distribution system [3], which is comprised by ten buses and nine lines (see Fig. 7.5). The first bus in the system is represented as the substation bus, which has a supplied voltage of 23.0 kV. The

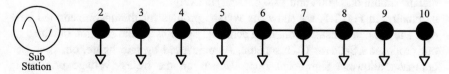

Fig. 7.5 IEEE's 10-bus distribution system

remaining nine buses are represented as load buses, all of which comprise a total active and reactive power load of 12,368.0 kW and 4186.0 kVAR, respectively.

For the 10-Bus distribution system under uncompensated conditions (this is, without installing any shunt capacitor on said system), it is considered that the annual power loss that results of its regular operation is of about 783.77 kW, which leads to a total annual cost of $131,674.00 and $411,949.51 according to both, case 1 and 2, respectively. Also, for said base case, the minimum and maximum voltages present on the buses are found to be of 0.8404 and 0.9930 p.u., respectively. More detailed data about the considered 10-Bus distribution system may be found in the Appendix section.

In Tables 7.4 and 7.5, the comparative results corresponding to our proposed approach and several other techniques (such as those based on Plant Growth Simulation Algorithm (PGSA) [33], Flower Pollination Algorithm (FPA) [34], Fuzzy Logic [35], Modified Cultural Algorithm [36], and Crow Search Algorithm (CSA) [3]) applied for OCP on the 10-Bus distribution system, are shown. From the results reported in Table 7.4 (which are given considering the minimization problem illustrated by Eq. 7.19 as the cost function) it could be appreciated our proposed approach has chosen buses 3, 4, 5, 6, 7, 9 and 10 as optimal buses to allocate shunt capacitors, in this case of the sizes 4050, 2250, 1950, 900, 450, 150 and 450 kVAR, respectively. Said capacitor allocation allows to reduce the total power loss to 675.2604 kW, which represents net savings of about a 12.29% in comparison to the uncompensated case. Also, the voltage profile on the buses also shown a significant improvement, achieving a minimum and maximum of 0.9024 and 1.007 p.u., respectively. When compared to the other considered OCP technique of our study, the solution provided by our proposed LS-Based approach leads to both, the most reduction on power loss and a better improvement on the buses' voltage profiles, and thus, shows the competitiveness our proposed approach. On the other hand, the results reported on Table 7.5 (which correspond those related to the minimization problem illustrated by Eq. 7.20) our proposed approach has selected the buses numbered 5, 6, 8, 10 as candidate buses, with their respective capacitor sizes being of 1500, 900, 350, 350 kVAR, respectively. The power loss after capacitor compensation is reduced to 680.42 kW, which combined with installation and operating costs generates net saving for a 9.95% percent in comparison to the uncompensated case. Although the total power loss generated by applying the solution provided by our LS-Based OCP approach is slightly greater in comparison to the other compared techniques, it should be noted that our approach still gets the greatest amount of annual net savings compared to the other methods. Finally, the improvement on voltage profiles that our proposed approach achieves should also be noted, with it having reached a minimum and maximum of 0.9004 and 1.00 p.u., respectively.

Finally, in Fig. 7.5, we show the voltage profiles for the uncompensated base case and the compensated condition achieved by applying the solution obtained by our proposed LS-Based OCP scheme. As evidenced by said figure, our proposed approach allows a significant improvement on the buses' voltage profiles in comparison to the uncompensated case (Fig. 7.6).

Table 7.4 Experimental results for the IEEE's 10-bus distribution system by considering the minimization problem illustrated by case 1

Concept	Base case	PGSA	FPA	Fuzzy logic	MCA	CSA	LS
Total power losses (kW)	783.77	694.93	688.28	686.00	677.54	676.22	675.26
Power losses total Cost ($)	131,673.36	116,748.24	115,631.04	115,248.00	113,826.216	113,604.12	113,443.74
Optimal Buses Index (No.)	–	6, 5, 9, 10	5, 7, 9, 10	3, 5, 6, 9	3, 4, 5, 6, 8, 10	3, 4, 5, 6, 7, 10	3, 4, 5, 6, 7, 9, 10
Optimal capacitor size (kVAR)	–	1200, 1200, 200, 407	1500, 300, 600, 1100	3600, 4050, 1650, 600	3300, 2850, 1950, 1200, 300, 450	4050, 2100, 1950, 900, 450, 600	4050, 2250, 1950, 900, 450, 150, 450
Total reactive power (kVAR)	–	3007	3500	9850	10,058	10,050	10,200
Capacitors cost ($)	–	1,591.76	789.30	1,787.40	1,930.05	1,916.55	2,047.10
Total annual cost ($)	131,673.36	118,340.00	116,420.34	117,035.40	115,756.26	115,520.67	115,490.84
Net saving ($)	–	13,334.00	15,253.02	14,637.96	15,917.1	16,152.69	16,183.00
Net savings (%)	–	10.13	11.58	11.11	12.08	12.25	12.29
Minimum V (p.u.)	0.8375	–	0.9509	0.9003	0.9001	–	0.9024
Maximum V (p.u.)	0.9929	–	–	1.007	1.007	–	1.007

Table 7.5 Experimental results for the IEEE's 10-bus distribution system by considering the minimization problem illustrated by Case 2

Concept	Base case	PGSA	FPA	Fuzzy logic	MCA	CSA	LS
Total power losses (kW)	783.77	694.93	688.28	686.00	677.54	676.22	680.42
Power losses total cost ($)	411,949.51	365,255.20	361,759.96	360,561.60	356,115.03	355,421.23	357,629.69
Optimal buses index (No.)	–	6, 5, 9, 10	5, 7, 9, 10	3, 5, 6, 9	3, 4, 5, 6, 8, 10	3, 4, 5, 6, 7, 10	5, 6, 8, 10
Optimal capacitor size (kVAR)	–	1200, 1200, 200, 407	1500, 300, 600, 1100	3600, 4050, 1650, 600	3300, 2850, 1950, 1200, 300, 450	4050, 2100, 1950, 900, 450, 600	1500, 900, 350, 350
Total reactive power (kVAR)	–	3,007	3500	9850	10,058	10,050	3100
Capacitors cost ($)	–	13,021.00	14,500.00	33,550.00	36,174.00	36,150.00	13,300.00
Total annual cost ($)	131,673.36	378,276.21	376,259.96	394,111.60	392,289.03	391,571.23	370,929.69
Net saving ($)	–	33,673.30	35,689.54	17,837.79	19,660.49	20,378.28	41,019.82
Net savings (%)	–	8.17	8.66	4.33	4.77	4.94	9.95
Minimum V (p.u.)	0.8375	–	0.9509	0.9003	0.9001	–	0.9004
Maximum V (p.u.)	0.9929	–	–	1.007	1.007	–	1.000

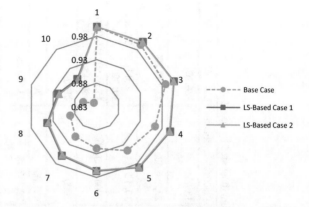

Fig. 7.6 Voltage profiles for the IEEEs 10-bus distribution system: uncompensated case VS LS-based compensated case (1 and 2)

7.5.3 Comparative Results for IEEE's 33-Bus Distribution System

For our second set of experiments the IEEE's 33-Bus distribution system has been considered [11, 21]. This electrical system is comprised by the 33 buses and 32 lines, with the first of said buses being considered as the substation bus (see Fig. 7.7). The voltage supplied by said substation is set as 12.66 kV, while the total active and reactive power provide by the remaining 32 (load) buses is of 3,715.00 kW and 2,300.00 kVAR, respectively.

Under uncompensated conditions, the system has a total power loss of about 210.78 kW, which leads to total annual costs of $35,442.96 and $110,885.83 with regard to cases 1 and 2, respectively. Also, the minimum and maximum voltages on the buses are found to be of 0.9037 and 0.9970 p.u., respectively. Further details about the analyzed 33-Bus distribution system may be found in the Appendix section.

In Tables 7.6 and 7.7, the comparative results corresponding to our proposed

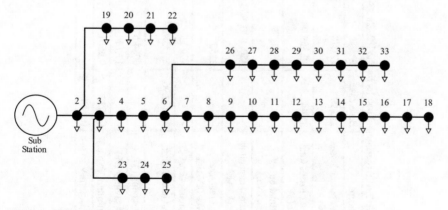

Fig. 7.7 IEEE's 33-bus distribution system

Table 7.6 Experimental results for the IEEE's 33-bus distribution system by considering the minimization problem illustrated by case 1

Concept	Base case	IP	SA	Two stage method	AA-FRCGA	Practical approach	LS
Total power losses (kW)	210.97	171.78	151.75	144.04	141.24	138.61	136.10
Power losses total cost ($)	35,442.96	28,859.04	25,494.00	24,199.06	23,727.61	23,286.48	22,865.20
Optimal buses index (No.)	–	9, 29, 30	10, 30, 14	7, 29, 30	28, 6, 29, 8, 30, 9	30, 12, 24	5, 8, 11, 16, 24, 26, 30, 32
Optimal capacitor size (kVAR)	–	450, 800, 900	450, 350, 900	850, 25, 900	25, 475, 300, 175, 400, 350	1000, 500, 500	150, 150, 150, 150, 450, 150, 750, 150
Total reactive power (kVAR)	–	2150	1700	1775	1725	2000	2100
capacitors cost ($)	–	499.35	401.05	507.15	492.86	481.00	770.85
Total annual cost ($)	35,442.96	29,358.39	25,895.05	24,706.21	24,220.05	23,767.48	23,636.05
Net saving ($)	–	6084.57	9,547.91	10,736.75	11,222.91	11,675.48	11,806.91
Net savings (%)	–	17.17	26.93	30.29	31.66	32.94	33.31
Minimum V (p.u.)	0.9037	0.9501	0.9591	0.9251	0.9665	–	0.9325
Maximum V (p.u.)	0.9970	–	–	–	–	–	0.9977

Table 7.7 Experimental results for the IEEE's 33-bus distribution system by considering the minimization problem illustrated by case 2

Concept	Base case	IP	SA	Two stage method	AA-FRCGA	Practical approach	LS
Total power losses (kW)	210.97	171.78	151.75	144.04	141.24	138.61	139.23
Power losses total cost ($)	110,885.83	90,287.56	79,759.80	75,707.42	74,235.74	72,853.41	73,179.28
Optimal busses index (No.)	–	9, 29, 30	10, 30, 14	7, 29, 30	28, 6, 29, 8, 30, 9	30, 12, 24	12, 25, 30
Optimal capacitor size (kVAR)	–	450, 800, 900	450, 350, 900	850, 25, 900	25, 475, 300, 175, 400, 350	1000, 500, 500	450, 350, 900
Total reactive power (kVAR)	–	2,150	1700	1775	1725	2000	1700
Capacitors cost ($)	–	9450	8100	8325	11,175	9000	8100
Total annual cost ($)	110,885.83	99,737.56	87,859.80	84,032.42	85,410.74	81,853.41	81,279.28
Net saving ($)	–	11,148.26	23,026.03	26,853.40	25,475.08	29,032.41	29,606.54
Net savings (%)	–	10.05	20.76	24.21	22.97	26.18	26.70
Minimum V (p.u.)	0.9037	0.9501	0.9591	0.9251	0.9665	–	0.9291
Maximum V (p.u.)	0.9970	–		–	–	–	1.0000

method and some other techniques reported on the literature for OCP applied to the IEEE's 33-Bus distribution system are shown. For said comparison, approaches such as those based on Interior Point (IP) [1], Simulated Annealing (SA) [1], the Two Stage Method [37], Analytical Algorithm and Fuzzy-Real Coded Genetic Algorithm (AA-FRCGA) [2] and the Practical Approach proposed in [38] have been considered. In Table 7.6, the results corresponding to the minimization problem presented in Eq. 7.19 are exposed. As shown by this table, the buses 5, 8, 11, 16, 24, 26, 30 and 32 has been chosen by our proposed LS-Based OCP approach as optimal buses to place the shunt capacitors, in this case of the sizes 150, 150, 150, 150, 450, 150, 750 and 150 kVAR, respectively. The reactive power compensation provided by such a configuration on capacitor placement allows to reduce the total power loss to 136.1028 kW, which yields to net savings of about a 33.31% in comparison to the uncompensated case. Furthermore, a slight improvement on the buses' voltage profile is also achieved, yielding to a minimum and maximum of 0.9325 p.u. and 0.9977, respectively. Also, a closer examination of Table 7.3 also shows that our proposed LS-Based OCP approach leads to the most reduction on power loss when compared to the other methods considered in our study, and as such, further demonstrate the competitiveness of our proposed method. As for the results reported in Table 7.7 (which correspond to the minimization problem illustrated in Eq. 7.20), it is shown that our proposed approach has chosen buses 12, 25 and 30 as candidate buses for placing shunt capacitors of sizes 450, 350 and 900 kVAR, respectively, leading to annual net savings of about a 26.70% when compared to the uncompensated case. In addition, it could be appreciated that the voltage profile has also improved, achieving a minimum and maximum of 0.9291 and 1.00 p.u., respectively. Finally, in Fig. 7.8 the voltage profiles for the uncompensated base case and the compensated condition achieved by applying the solution obtained by our proposed LS-Based OCP method are shown. Similarly, to the case studied the previous section, there is an evident improvement on the buses' voltage profiles when compared to the uncompensated case.

Fig. 7.8 Voltage profiles for the IEEEs 33-bus distribution system: uncompensated case VS LS-based compensated case (1 and 2)

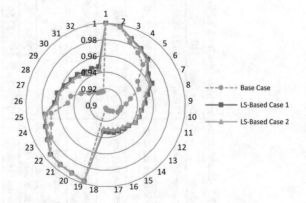

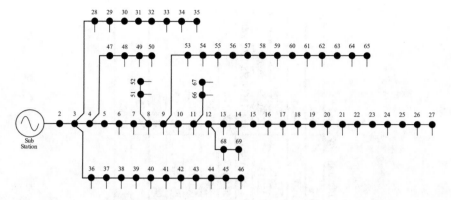

Fig. 7.9 IEEE's 69-bus distribution system

7.5.4   Comparative Results for IEEE's 69-Bus Distribution System

For our last set of experiments, we choose the IEEE's 69-Bus distribution system [11, 21] (see Fig. 7.9). In this case, the voltage supplied by the substation bus (bus one) is of 12.66 kV, while the total active and reactive power loads are found to be of 3801.89 kW and 2693.60 kVAR, respectively.

Under uncompensated conditions, the distribution system has a total power loss of about 225 kW, which leads to a total annual cost of $37,800.00 and $118,260.00 with regard to cases 1 and 2, respectively. Also, the minimum and maximum voltages for the system's buses are found to be of 0.9091 and 0.9999 p.u., respectively. Finally, similarly the previous two cases, the costs related to installation and maintenance are not considered. Further details about the 69-Bus distribution system are given on the Appendix section.

In Tables 7.8 and 7.9, the comparative results corresponding to the LS-Based OCP method and several other techniques reported on the literature to solve the 69-Bus distribution system OCP, such as those based on Flower Pollination Algorithm (FPA) [39], Direct Search Algorithm (DSA) [40], Teaching Learning Based Optimization (TLBO) [41], Crow Search Algorithm (CSA) [3] and Gravitational Search Algorithm (GSA) [1], are presented. In Table 7.8, the results corresponding to the minimization problem established in Eq. 7.19 are shown. From this table, it could be appreciated that our proposed method has found that the buses indexed as 12, 21, 50, 54 and 61 are optimal for placing shunt capacitors, with their corresponding sizes being 350, 150, 450, 150, and 1200 kVAR, respectively.

By considering said capacitor placement configuration, the total power loss is reduced from 225 kW to 144.2552 kW, which represents net savings of about a 34.34% over the total annual cost under uncompensated conditions. Furthermore, the minimum and maximum voltage on the buses also suffers a slight improvement, with those reaching values of 0.9315 and 1.0000 p.u., respectively. As with the experiment

Table 7.8 Experimental results for the IEEE's 69-bus distribution system by considering the minimization problem illustrated by case 1

Concept	Base case	Base case	FPA	DSA	TLBO	CSA	GSA
Total power losses (kW)	225.00	150.28	147.00	146.35	146.10	145.90	144.25
Power losses total cost ($)	37,800.00	25,247.04	24,696.00	24,586.80	24,544.80	24,511.20	24,234.87
Optimal buses index (No.)	–	61	61, 15, 60	12, 61, 64	18, 61, 65	26, 13, 15	12, 21, 50, 54, 61
Optimal capacitor size (kVAR)	–	1350	900, 450, 450	600, 1050, 150	350, 1150, 150	150, 150, 1050	350, 150, 450, 150, 1200
Total reactive power (kVAR)	–	1,350.00	1,800.00	1,800.00	1,650.00	1,350.00	2,300.00
Capacitors cost ($)	–	279.45	392.40	446.40	459.70	451.50	590.35
Total annual cost ($)	37,800.00	25,526.49	25,088.40	25,033.20	25,004.50	24,962.70	24,825.22
Net saving ($)	–	12,273.51	12,711.60	12,766.80	12,795.50	12,837.30	12,974.78
Net savings (%)	–	32.46	33.62	33.77	33.85	33.94	34.32
Minimum V (p.u.)	0.90901	–	–	0.9313	0.9321	–	0.9315
Maximum V (p.u.)	0.9999	–	–	–	–	–	1.0000

Table 7.9 Experimental results for the IEEE's 69-bus distribution system by considering the minimization problem illustrated by case 2

Concept	Base case	Base case	FPA	DSA	TLBO	CSA	GSA
Total power losses (kW)	225.00	150.28	147.00	146.35	146.10	145.90	146.61
Power losses total cost ($)	118,260.00	78,987.16	77,263.20	76,921.56	76,790.16	76,685.04	77,058.22
Optimal buses index (No.)	–	61	61, 15, 60	12, 61, 64	18, 61, 65	26, 13, 15	17, 61
Optimal capacitor size (kVAR)	–	1350	900, 450, 450	600, 1050, 150	350, 1150, 150	150, 150, 1050	350, 1200
Total reactive power (kVAR)	–	1350.00	1,800.00	1,800.00	1,650.00	1,350.00	1,550.00
Capacitors cost ($)	–	5050	8400	8400	7950	7050	6650
Total annual cost ($)	118,260.00	84,037.16	85,663.20	85,321.56	84,740.16	83,735.04	83,708.21
Net saving ($)	–	34,222.83	32,596.80	32,938.44	33,519.84	34,524.96	34,551.78
Net savings (%)	–	28.94	27.56	27.85	28.34	29.19	29.21
Minimum V (p.u.)	0.90901	–	–	0.9313	0.9321	–	0.9300
Maximum V (p.u.)	0.9999	–	–	–	–	–	1.0000

Fig. 7.10 Voltage profiles
for the IEEEs 69-bus
distribution system:
uncompensated case VS
LS-based compensated case
(1 and 2)

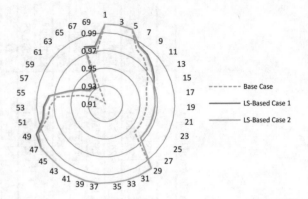

reported on the previous two sections, Table 7.8 also shows that the proposed LS-Based OCP scheme provides the best OCP configuration in comparison to the other compared techniques, as it allows to most reduction on power losses. This once again proves the competitiveness of the proposed LS-Based approach when applied to the complex OCP task. Furthermore, in Table 7.9, the cost minimization results that arise from considering the minimization problem illustrated by Eq. 7.20 are presented.

In this case, our proposed LS-Based OCP approach consider buses number 17 and 16 as candidate locations for applying reactive compensation, with the capacitors to be placed being of the values 350 and 1200 kVAR, respectively. With this capacitor configuration annual net savings of about a 29.21% in comparison to the uncompensated case are achieved. Furthermore, it should also be appreciated that the voltage profile also presents a significant improvement over that of the base case, with it reaching minimum and maximum values of 0.9300 and 1.00 p.u., respectively. Finally, Fig. 7.10 shows the voltage profiles for both the uncompensated case and the compensated condition achieved by applying the OCP configuration proposed by our LS-Based method. As with the two distribution network cases studied in the previous sections, significant improvements on the buses' voltage profiles can be appreciated when compared to the uncompensated case.

7.6   Conclusions

In this work, the swarm optimization algorithm known as LS has been proposed to solve the high multi-modal OCP problem on RDNs. Different to other swarm schemes, the LS algorithm explicitly avoids the concentration of individuals around the best solutions during the search process. This important characteristic allows to solve complex high multi-modal formulations with discontinuities and nonlinearities. The proposed scheme has been able to find highly competitive solutions, yielding both, a notorious reduction in power loss and an improvement of voltage profiles on RDNs. In order to demonstrate the performance of the proposed LS-Based OCP scheme, it has been tested by considering several RDN models commonly reported on the literature as reference, specifically the IEEE's 10-Bus, 33-Bus and 69-Bus distribution systems. For each of the analyzed distribution systems, our proposed OCP method has been also compared against other OCP techniques currently reported on the literature. The results demonstrate the high performance of the proposed method in terms of accuracy and robustness.

Appendix

In Tables 7.10, 7.11 and 7.12 the line data corresponding to the IEEE's 10-Bus, 33-Bus and 69-Bus Distribution Systems, as presented in [3, 16, 17], is shown.

Table 7.10 IEEE's 10-bus distribution system line data

Line No.	Line bus index i	Line bus index i + 1	Line resistance R (Ω)	Line reactance X (Ω)	Real load power PL (kW)	Reactive load power QL (kVAR)
1	1	2	0.1233	0.4127	1840	460
2	2	3	0.0140	0.6051	980	340
3	3	4	0.7463	1.2050	1790	446
4	4	5	0.6984	0.6084	1598	1840
5	5	6	1.9831	1.7276	1610	600
6	6	7	0.9053	0.7886	780	110
7	7	8	2.0552	1.1640	1150	60
8	8	9	4.7953	2.7160	980	130
9	9	10	5.3434	3.0264	1640	200

Table 7.11 IEEE's 33-bus distribution system line data

Line No.	Line bus index i	Line bus index i + 1	Line resistance R (Ω)	Line reactance X (Ω)	Real load power PL (kW)	Reactive load power QL (kVAR)
1	1	2	0.0922	0.0477	100	60
2	2	3	0.4930	0.2511	90	40
3	3	4	0.3660	0.1864	120	80
4	4	5	0.3811	0.1941	60	30
5	5	6	0.8190	0.7070	60	20
6	6	7	0.1872	0.6188	200	100
7	7	8	1.7114	1.2351	200	100
8	8	9	1.0300	0.7400	60	20

(continued)

Table 7.11 (continued)

Line No.	Line bus index i	Line bus index i + 1	Line resistance R (Ω)	Line reactance X (Ω)	Real load power PL (kW)	Reactive load power QL (kVAR)
9	9	10	1.0400	0.7400	60	20
10	10	11	0.1966	0.0650	45	30
11	11	12	0.3744	0.1238	60	35
12	12	13	1.4680	1.1550	60	35
13	13	14	0.5416	0.7129	120	80
14	14	15	0.5910	0.5260	60	10
15	15	16	0.7463	0.5450	60	20
16	16	17	1.2890	1.7210	60	20
17	17	18	0.7320	0.5740	90	40
18	2	19	0.1640	0.1565	90	40
19	19	20	1.5042	1.3554	90	40
20	20	21	0.4095	0.4784	90	40
21	21	22	0.7089	0.9373	90	40
22	3	23	0.4512	0.3083	90	50
23	23	24	0.8980	0.7091	420	200
24	24	25	0.8960	0.7011	420	200
25	6	26	0.2030	0.1034	60	25
26	26	27	0.2842	0.1447	60	25
27	27	28	1.0590	0.9337	60	20
28	28	29	0.8042	0.7006	120	70
29	29	30	0.5075	0.2585	200	600
30	30	31	0.9744	0.9630	150	70
31	31	32	0.3105	0.3619	210	100
32	32	33	0.3410	0.5302	60	40

Table 7.12 IEEE's 69-bus distribution system line data

Line No.	Line bus index i	Line bus index i + 1	Line resistance R (Ω)	Line reactanc X (Ω)	Real load power PL (kW)	Reactive load power QL (kVAR)
1	1	2	0.00050	0.0012	0.00	0.00
2	2	3	0.00050	0.0012	0.00	0.00
3	3	4	0.00150	0.0036	0.00	0.00
4	4	5	0.02510	0.0294	0.00	0.00
5	5	6	0.36600	0.1864	2.60	2.20
6	6	7	0.38100	0.1941	40.40	30.00
7	7	8	0.09220	0.0470	75.00	54.00
8	8	9	0.04930	0.0251	30.00	22.00
9	9	10	0.81900	0.2707	28.00	19.00
10	10	11	0.18720	0.0619	145.00	104.00
11	11	12	0.71140	0.2351	145.00	104.00
12	12	13	1.03000	0.3400	8.00	5.00
13	13	14	1.04400	0.3400	8.00	5.00
14	14	15	1.05800	0.3496	0.00	0.00
15	15	16	0.19660	0.0650	45.00	30.00
16	16	17	0.37440	0.1238	60.00	35.00
17	17	18	0.00470	0.0016	60.00	35.00
18	18	19	0.32760	0.1083	0.00	0.00
19	19	20	0.21060	0.0690	1.00	0.60
20	20	21	0.34160	0.1129	114.00	81.00
21	21	22	0.01400	0.0046	5.00	3.50
22	22	23	0.15910	0.0526	0.00	0.00
23	23	24	0.34630	0.1145	28.00	20.00
24	24	25	0.74880	0.2475	0.00	0.00
25	25	26	0.30890	0.1021	14.00	10.00
26	26	27	0.17320	0.0572	14.00	10.00
27	3	28	0.00440	0.0108	26.00	18.60
28	28	29	0.06400	0.1565	26.00	18.60
29	29	30	0.39780	0.1315	0.00	0.00
30	30	31	0.07020	0.0232	0.00	0.00
31	31	32	0.35100	0.1160	0.00	0.00
32	32	33	0.83900	0.2816	14.00	10.00
33	33	34	1.70800	0.5646	19.50	14.00
34	34	35	1.47400	0.4873	6.00	4.00
35	3	36	0.00440	0.0108	26.00	18.55

(continued)

Table 7.12 (continued)

Line No.	Line bus index i	Line bus index i + 1	Line resistance R (Ω)	Line reactanc X (Ω)	Real load power PL (kW)	Reactive load power QL (kVAR)
36	36	37	0.06400	0.1565	26.00	18.55
37	37	38	0.10530	0.1230	0.00	0.00
38	38	39	0.03040	0.0355	24.00	17.00
39	39	40	0.00180	0.0021	24.00	17.00
40	40	41	0.72830	0.8509	1.20	1.00
41	41	42	0.31000	0.3623	0.00	0.00
42	42	43	0.04100	0.0478	6.00	4.30
43	43	44	0.00920	0.0116	0.00	0.00
44	44	45	0.10890	0.1373	39.22	26.30
45	45	46	0.00090	0.0012	39.22	26.30
46	4	47	0.00340	0.0084	0.00	0.00
47	47	48	0.08510	0.2083	79.00	56.40
48	48	49	0.28980	0.7091	384.70	274.50
49	49	50	0.08220	0.2011	384.70	274.50
50	8	51	0.09280	0.0473	40.50	28.30
51	51	52	0.33190	0.1140	3.60	2.70
52	9	53	0.17400	0.0886	4.35	3.50
53	53	54	0.20300	0.1034	26.40	19.00
54	54	55	0.28420	0.1447	24.00	17.20
55	55	56	0.28130	0.1433	0.00	0.00
56	56	57	1.59000	0.5337	0.00	0.00
57	57	58	0.78370	0.2630	0.00	0.00
58	58	59	0.30420	0.1006	100.00	72.00
59	59	60	0.38610	0.1172	0.00	0.00
60	60	61	0.50750	0.2585	1244.00	888.00
61	61	62	0.09740	0.0496	32.00	23.00
62	62	63	0.14500	0.0738	0.00	0.00
63	63	64	0.71050	0.3619	227.00	162.00
64	64	65	1.04100	0.5302	59.00	42.00
65	11	66	0.20120	0.0611	18.00	13.00
66	66	67	0.00470	0.0014	18.00	13.00
67	12	68	0.73940	0.2444	28.00	20.00
68	68	69	0.00470	0.0016	28.00	20.00

References

1. Mohamed Shuaib, Y., Surya Kalavathi, M., Christober Asir Rajan, C.: Optimal capacitor placement in radial distribution system using gravitational search algorithm. Int. J. Electr. Power Energy Syst. **64**, 384–397 (2015)
2. Abul'Wafa, A.R.: Optimal capacitor allocation in radial distribution systems for loss reduction: A two stage method. Electr. Power Syst. Res. **95**, 168–174 (2013)
3. Askarzadeh, A.: Capacitor placement in distribution systems for power loss reduction and voltage improvement: a new methodology. IET Gener. Transm. Distrib. **10**(14), 3631–3638 (2016)
4. Aman, M.M., Jasmon, G.B., Bakar, A.H.A., Mokhlis, H., Karimi, M.: Optimum shunt capacitor placement in distribution system—A review and comparative study. Renew. Sustain. Energy Rev. **30**, 429–439 (2014)
5. Ng, H.N., Salama, M.M.A., Chikhani, A.Y.: Classification of capacitor allocation techniques. IEEE Trans. Power Deliv. **15**(1), 387–392 (2000)
6. Neagle, N.M., Samson, D.R.: Loss reduction from capacitors installed on primary feeders [includes discussion]. Trans. Am. Inst. Electr. Eng. Part III Power Appar. Syst. **75**(3), 950–959 (1956)
7. Cook, R.F.: Analysis of capacitor application as affected by load cycle. Trans. Am. Inst. Electr. Eng. **78**, 950–957 (1959)
8. Cook, R.F.: Optimizing the application of shunt capacitors for reactive-volt-ampere control and loss reduction. Trans. Am. Inst. Electr. Eng. Part III Power Appar. Syst. **80**(3), 430–441 (1961)
9. Schmill, J.V.: Optimum size and location of shunt capacitors on distribution feeders. IEEE Trans. Power Appar. Syst. **84**(9), 825–832 (1965)
10. Bae, Y.G.: Analytical method of capacitor allocation on distribution primary feeders. IEEE Trans. Power Appar. Syst. **PAS-97**(4), 1232–1238 (1978)
11. Baran, M.E., Wu, F.F.: Optimal capacitor placement on radial distribution systems. IEEE Trans. Power Deliv. **4**(1), 725–734 (1989)
12. Durán, H.: Optimum number, location, and size of shunt capacitors in radial distribution feeders a dynamic programming approach. IEEE Trans. Power Appar. Syst. **PAS-87**(9), 1769–1774 (1968)
13. Pannavaikko, M., Prakasa Rao, K.S.: Optimal choice of fixed and switched shunt capacitors on radial distributors by the method of local variations. IEEE Trans. Power Appar. Syst. **102**(6), 1607–1615 (1983)
14. Cuevas, E., Osuna, V., Oliva, D.: Evolutionary computation techniques: a comparative perspective **686** (2017)
15. Cuevas, E., Díaz Cortés, M.A., Oliva Navarro, D.A.: Advances of Evolutionary Computation: Methods and Operators, 1st ed. Springer International Publishing, Berlin (2016)
16. Singh, S.P., Rao, A.R.: Optimal allocation of capacitors in distribution systems using particle swarm optimization. Int. J. Electr. Power Energy Syst. **43**(1), 1267–1275 (2012)
17. Devabalaji, K.R., Ravi, K., Kothari, D.P.: Optimal location and sizing of capacitor placement in radial distribution system using bacterial foraging optimization algorithm. Int. J. Electr. Power Energy Syst. **71**, 383–390 (2015)
18. Díaz, P., et al.: An improved crow search algorithm applied to energy problems, pp. 1–23 (2018)
19. Díaz-Cortés, M.-A., Cuevas, E., Rojas, R.: Engineering Applications of Soft Computing (2017)
20. Cuevas, E., González, A., Zaldívar, D., Cisneros, M.P.: An optimisation algorithm based on the behaviour of locust swarms. Int. J. Bio-Inspired Comput. **7**(6), 402 (2015)
21. Baran, M.E., Wu, F.F.: Network reconfiguration in distribution systems for loss reduction and load balancing. Power Deliv. IEEE Trans. **4**(2), 1401–1407 (1989)
22. Flaih, F.M.F., Xiangning, L., Dawoud, S.M.: Distribution system reconfiguration for power loss minimization and voltage profile improvement using Modified particle swarm optimization. Power Energy, pp. 120–124 (2016)

23. C.-T. Su and C.-C. Tsai, "A new fuzzy-reasoning approach to optimum capacitor allocation for primary distribution systems," *Proc. IEEE Int. Conf. Ind. Technol.*, pp. 237–241, 1996
24. El-Fergany, A.A.: Optimal capacitor allocations using integrated evolutionary algorithms. Int. Rev. Model. Simulations **5**(6), 2590–2599 (2012)
25. Prakash, D.B., Lakshminarayana, C.: Optimal siting of capacitors in radial distribution network using Whale optimization algorithm. Alexandria Eng. J. **56**(4), 499–509 (2017)
26. Cuevas, E.: Block-matching algorithm based on harmony search optimization for motion estimation. Appl. Intell. **39**(1), 165–183 (2013)
27. Díaz-Cortés, M.-A., Ortega-Sánchez, N., Hinojosa, S., Cuevas, E., Rojas, R., Demin, A.: A multi-level thresholding method for breast thermograms analysis using Dragonfly algorithm. Infrared Phys. Technol. **93**, 346–361 (2018)
28. Díaz, P., Pérez-Cisneros, M., Cuevas, E., Hinojosa, S., Zaldivar, D.: An im-proved crow search algorithm applied to energy problems. Energies **11**(3), 571 (2018)
29. González, A., Cuevas, E., Fausto, F., Valdivia, A., Rojas, R.: A template matching approach based on the behavior of swarms of locust. Appl. Intell. **47**(4), 1087–1098 (2017)
30. Laskari, E.C., Parsopoulos, K.E., Vrahatis, M.N.: Particle swarm optimization for integer programming. In: Proceedings of the 2002 Congress Evaluation Computing CEC'02 (Cat. No.02TH8600), vol. 2, pp. 1582–1587 (2002)
31. Mekhamer, S.F., Soliman, S., Moustafa, M., El-Hawary, M.E.: Application of fuzzy logic for reactive-power compensation of radial distribution feeders. IEEE Trans. Power Syst. **18**(1), 206–213 (2003)
32. Baran, M.E., Wu, F.F.: Optimal sizing of capacitors placed on a radial distribution system. IEEE Trans. Power Deliv. **4**(1), 735–743 (1989)
33. Rao, R.S., Narasimham, S.V.L., Ramalingaraju, M.: Optimal capacitor placement in a radial distribution system using Plant Growth Simulation Algorithm. Int. J. Electr. Power Energy Syst. **33**(5), 1133–1139 (2011)
34. Abdelaziz, A.Y., Ali, E.S., Abd Elazim, S.M.: Flower pollination algorithm for optimal capacitor placement and sizing in distribution systems. Electr. Power Components Syst. **44**(5), 544–555 (2016)
35. Seifi, A.R.: A new hybrid optimization method for optimum distribution capacitor planning **2**, 819–824 (2008)
36. Haldar, V., Chakraborty, N.: Power loss minimization by optimal capacitor placement in radial distribution system using modified cultural algorithm. Int. Trans. Electr. ENERGY Syst. **25**, 54–71 (2015)
37. Abul'Wafa, A.R.: Optimal capacitor placement for enhancing voltage stability in distribution systems using analytical algorithm and Fuzzy-Real Coded GA. Int. J. Electr. Power Energy Syst. **55**, 246–252 (20142014)
38. Hung, D.Q., Mithulananthan, N., Bansal, R.C.: A combined practical approach for distribution system loss reduction. Int. J. Ambient Energy **36**(3), 123–131 (2015)
39. Abdelaziz, A.Y., Ali, E.S., Abd Elazim, S.M.: Flower pollination algorithm and loss sensitivity factors for optimal sizing and placement of capacitors in radial distribution systems. Int. J. Electr. Power Energy Syst. **78**, 207–214 (2016)
40. Ramalinga Raju, M., Ramachandra Murthy, K.V.S., Ravindra, K.: Direct search algorithm for capacitive compensation in radial distribution systems. Int. J. Electr. Power Energy Syst. **42**(1), 24–30 (2012)
41. Sultana, S., Roy, P.K.: Optimal capacitor placement in radial distribution systems using teaching learning based optimization. Int. J. Electr. Power Energy Syst. **54**, 387–398 (2014)

Chapter 8
Blood Vessel and Optic Disc Segmentation Based on a Metaheuristic Method

Abstract In recent years, image processing techniques have been an essential tool for health care. A good and timely analysis of retinal vessel images has become relevant in the identification and treatment of diverse cardiovascular and ophthalmological illness. Therefore, an automatic and precise method for retinal vessel and optic disc segmentation is crucial for illness detection. This task is arduous, time-consuming, and generally developed by an expert with a considerable grade of professional skills in the field. Various retinal vessel segmentation approaches have been developed with promissory results. Although, most of such methods present a deficient performance principally due to the complex structure of vessels in retinal images. In this work, an accurate and hybrid methodology for retinal vessel and optic disc segmentation is presented. The method proposed a fusion of two different schemas: the lateral inhibition (LI) and Differential Evolution (DE). LI is used to improve the contrast between the retinal vessel and background. Followed by the minimization of the cross-entropy function to find the threshold value, these performed by the second schema the DE algorithm. To test the performance and accuracy of the proposed methodology, a set of images obtained from three public datasets STARE, DRIVE, and DRISHTI-GS have used in different experiments. Simulation results demonstrate the high performance of the proposed approach in comparison with related methods reported in the literature.

8.1 Introduction

The retinal vessel patter represents an interesting information resource for medical diagnosis. The modification in its principal components such as length, width, angles, and branching structure, may imply the existence of some disease, for instance, hypertension, diabetes, choroidal neovascularization, arteriosclerosis, stroke, and cardiovascular problems [1–4]. The appearance of a distortion in the width of blood vessels due to capillary pressurizations is known as microaneurysms, and it is the primary signal of Diabetic Retinopathy, which is one of the most common causes of loss of sight in humans [5]. Therefore, an automatic and precise method for

© Springer Nature Switzerland AG 2021
E. Cuevas et al., *Metaheuristic Computation: A Performance Perspective*,
Intelligent Systems Reference Library 195,
https://doi.org/10.1007/978-3-030-58100-8_8

retinal vessel and optic disc segmentation has become a crucial task for assisting on the diagnostic of a possible retinal illness. In early days the segmentation was made manually which is hard and exhausting, and inefficient work. An alternative is to automatize the segmentation process. However, an automatic retinal vessel segmentation is not a trivial task due to the other components in the eye, such as optic disk, fovea, macula, etc. Additionally, the complexity is increased by the wide variety of width in the bifurcations of the retinal vessels. Under such conditions, some vessels are too narrow to be properly distinguished from the eye background.

To implement the automatically segmentation of blood vessel many alternatives have been proposed. The methodologies are grouped into different classes according on the approach they are based on. Instances of these categories represent areas such as pattern recognition, supervised and unsupervised machine learning methods, tracking-based, model-based, mathematical morphology, matched filtering, and multiscale approaches [6, 7]. Some examples include the works proposed in [1] and [8], where the authors introduced a supervised segmentation method based on a neural network. In [9], two convolutional neural networks are used as a feature extractor. Then, both are combined with the random forest method to classify the pixels in the image. In [8], the authors proposed the use of Gabor filters and moment invariants-based features to describe vessel and non-vessel information with the purpose of training a neural network for pixel classification. The authors in [10] proposed to segmentation method based on the combination of different techniques. Initially, the approach produces a new image with an enhancement of the non-uniform illumination from the green and red color channels. Then, a match filter is used to improve the contrast between background and blood vessels. Finally, a fuzzy c-means cluster algorithm is applied to classify the vessels pixels. In [11], self-organized maps are considered to produce a set of training samples from the original image. Afterward, a K-means algorithm is implemented to separate blood vessel elements from background pixels.

Considering a different scheme, in [12], the authors proposed a probabilistic tracking method based on a Bayesian classifier for segmenting blood vessel pixels. In the same way, other blood vessel segmentation method reported in [13] also combines a probabilistic tracking method with a multi-scale line detection scheme. Other interesting technique is proposed in [14] where blood vessel elements are segmented through the use of orthogonal projections of their texture information. In [15], it is proposed to divide the segmentation process into two steps. In the first stage, an image enhancement operation is implemented to remove noise, low contrast, and non-uniform illumination. In the second step, a morphological processing operation is used to classify blood vessel pixels. One approach reported in [16] considers a morphological operation along with a K-means algorithm to segment retinal vessels. In [17], the authors introduced a segmentation method that combines two elements. In the approach, a technique called Contrast Limited Adaptive Histogram Equalization (CLAHE) is considered to enhance the eyes image, while a binarization process is executed by combining Gaussian and Laplacian filters (LoG). In [18], a complete segmentation methodology is proposed. In the scheme, a matched filter based on a Gaussian function with zero-mean is implemented to detect vessels. Then, a threshold value is computed depending on the first-order derivate of the produced image. Authors in [19] presented a segmentation algorithm that applies an anisotropic diffusion filter to remove noise and reconnect vessel lines. Then, a multi-scale line tracking is used to identify vessels with similar size.

Recently, several new schemes for segmenting blood vessel and optic disc images have been developed through the use of machine learning techniques. Some examples of these approaches include a deep learning system [20] where a DirectNet architecture composed of blocks with Convolutional Neural Networks is implemented for vessel segmentation of retinal images. Another approach that uses Convolutional Neural Networks was also proposed in [21]. Here, authors face width and direction complexity of the retinal vessel structure by combining the multiscale analysis provided by the stationary wavelet transform with a multiscale Fully Convolutional Neural Network. In [22], a generalized method for blood vasculature detection has been proposed. This approach implements several steps in the segmentation process such as a preprocessing retinal fundus images for quality improvement, finding vascular and non-vascular structures to generate a vasculature map, the extraction of geometrical and intensity features from the vasculature map and the original fundus image, supervised classification of vascular and non-vascular structures, and finally a connectivity test to identify vascular structures. Authors in [23] proposed a technique were morphological processing and matched filtering are combined for detecting venule structure and capillaries. In [24], a framework for retinal image authentication has been proposed. This approach extracts characteristics from retinal images in order to identify features such as uniqueness and stability for authentication in security applications. The framework is based on a graph-based representation algorithm that faces rotation and intensity variations. Another retinal blood vessel segmentation has been proposed in [25]. In this work, vessels are segmented from fundus images through a pre-processing stage where several operations are applied to enhance the image. Then, supervised and unsupervised learning such as principal components analysis, clustering, and classification are applied to identify vessel structures. Finally, image post-processing techniques such as morphological operations take place to improve image segmentation. In [26], a supervised method that combines different features from different algorithms into a mix feature vector for pixel characterization is proposed. Then, a random forest classifier is trained with this feature vector for identifying vessel and non-vessel pixels in retinal images. Author in [27] proposed a multi-scale tensor voting approach for small retinal vessel segmentation. This technique focuses on the smallest vessel in fundus images. Here, line detection and perceptual organization methods are combined in a multi-scale scheme for reconstructing small vessels by tracking and pixel painting from the perceptual-based approach. In [28], an ensemble learning approach is also suggested for automatic disc segmentation.

Other recent interesting approaches have been proposed to automatically localize an extract retinal anatomical structures based on the combination of several schemes extracted of different computational fields. They include the fusion of adaptive fuzzy thresholding with mathematical morphology [29] to produce a hybrid segmentation algorithm for retinal vessel, optic disc, and exudate lesions extraction. In [30], a level-set approach is considered, while in [31] a Superpixel Classification method Based Optic Disc and Optic Cup is used for segmentation in Glaucoma Screening. In [32], a shape regression scheme is employed to segment optic disc and optic cup in retinal fundus images. On the other hand, in [33], methods of mathematical morphology have been considered for segmentation of fundus eye images. Another interesting scheme is [34] where several computer vision techniques are applied to measure the ocular torsion by using digital fundus image. Finally, in [35], a method based on a set of finite impulse response (FIR) filters is introduced for segmenting the optic disc.

Although most of these schemes have yielded notable results, they often present several difficulties due to the noisy, incomplete and imprecise nature of the retinal vessel images. Under such conditions, there still exist many critical problems that need to be addressed for recent research such as the presence of false positives, poor connectivity in retinal vessels, accuracy in noise conditions, and others.

Classical image processing techniques regularly face complications when they operate over images with vague or deficient information. In consequence, the use of metaheuristic optimization algorithms has been extensively studied to solve a wide range of computer vision problems such as circle detection [36], segmentation of blood cell and brain images [37, 38], template matching [39], and others. In general, metaheuristic methods have exhibited better results than those based on classical schemes in terms of robustness and accuracy [40–44]. The Differential Evolution (DE) algorithm [45] is one of the most popular metaheuristic techniques according to the literature. Different from other metaheuristic schemes, the search strategy of DE maintains interesting characteristics of exploration and exploitation without premature convergence [36]. Such abilities have promoted its use in a wide variety of applications [46–49].

In this work, an accurate methodology for retinal vessel and optic disc segmentation is presented. The proposed scheme combines two different techniques: the Lateral Inhibition (LI) and the Differential Evolution (DE). The LI scheme produces a new image with enhanced contrast between the background and retinal vessels. Then, the DE algorithm is used to obtain an appropriate threshold value through the minimization of the cross-entropy function from the enhanced image. Finally, morphological operations are applied to the final image for eliminating noise and artifacts in the output image. To evaluate the performance of the proposed approach, several experiments over images extracted from STARE, DRIVE, and DRISHTI-GS databases have been conducted. Simulation results demonstrate a high performance of the proposed scheme in comparison with similar methods reported in the literature.

The chapter is organized as follows: In Sect. 8.2 the main concepts used in the proposed scheme are explained. Section 8.3 presents the proposed methodology in detail. In Sect. 8.4, the experimental results and comparisons are exhibited and discussed. Finally, in Sect. 8.5, the conclusions are drawn.

8.2 Preliminary Concepts

In this section, the most important concepts used in our methodology is reviewed. They include lateral inhibition, cross entropy and differential evolution algorithm.

8.2.1 Lateral Inhibition

Lateral Inhibition (LI) [50, 51] is a neurobiological mechanism which involves the ability of an active receptor to decrease the activity of its neighbors. This effect generates a contrast in stimulation that increases the intensity of the perception [52]. Under LI, the closer proximity to an active receptor is, the more intense the inhibition effect would be [53]. In the context of our applications, receptors are pixels in the image, while their grayscale values represent their activation levels. With the LI principle, the objective is to increase the pixel differences in regions of high contrast whereas homogenous regions remain unaltered (4a). This operation is implemented by using a linear model [54] where the original grayscale image is denoted as I_g and the enhanced image is denoted as I_R. Therefore, the LI operator is conducted through the following formulation:

$$I_R(x, y) = I_g(x, y) + \sum_{i=-M}^{M} \sum_{j=-N}^{N} \delta_{ij} \cdot I_g(x + i, y + j) \tag{8.1}$$

where δ_{ij} corresponds to the LI coefficients. M and N specifies the size of the receptive field. Consequently, a grid schema with the values of $M = 2$ and $N = 2$ generates a structure of size 5×5. To produce the desired effect, the LI coefficients δ_{ij} must satisfy the following restriction:

$$\sum_{i=-M}^{M} \sum_{j=-N}^{N} \delta_{ij} = 0 \tag{8.2}$$

which implies balanced inhibition energy.

8.2.2 Cross Entropy

The cross-entropy [55] corresponds to the distance D between two probability distributions U and V. Assuming $U = \{u_1, u_2, \ldots, u_N\}$ and $V = \{v_1, v_2, \ldots, v_N\}$ as two discrete probability distributions, the cross entropy is defined as follow:

$$D(U, V) = \sum_{i=1}^{N} u_i \log \frac{u_i}{v_i} \tag{8.3}$$

Considering I_g as the original image and $h^g(i)$ as its corresponding histogram $h^g(i)$, the segmented image I_{th} is modeled as follows:

$$I_{th} = \begin{cases} \mu(1, t) & I_g(x, y) < t \\ \mu(t, L+1) & I_g(x, y) \geq t \end{cases},$$

$$i = 1, 2, \ldots, L \tag{8.4}$$

where t corresponds to the threshold that divides each pixel as background or object, while μ is defined in Eq. (8.8).

$$\mu(a, b) = \sum_{i=a}^{b-1} i h^g(i) / \sum_{i=a}^{b-1} h^g(i) \tag{8.5}$$

Therefore, the cross-entropy objective function D_{CE} [56] is formulated as follows:

$$D_{CE}(t) = - \sum_{i=1}^{t-1} i h^g(i) \log\left(\frac{1}{\mu(1, t)}\right) + \sum_{i=t}^{t-1} i h^g(i) \log\left(\frac{1}{\mu(t, L+1)}\right) \tag{8.6}$$

Under such conditions, the optimal threshold $\hat{t}$ is obtained by minimizing D_{CE}:

$$\hat{t} = \arg \min_t D_{CE}(t) \tag{8.7}$$

8.2.3   Differential Evolution Algorithm

Metaheuristic methods represents powerful tools that have demonstrated to solve complex engineering problems [57–59]. Differential Evolution (DE) [60] is a vector-based metaheuristic algorithm approach introduced by Storn and Price in 1996. Different from other metaheuristic methods [61], DE is one of the most simple and powerful optimization methods inspired by the phenomenon of evolution.

During its operation, in each generation s, DE applies a series of crossover, selection and mutation operators to allow the evolution of the population toward an optimal solution. This population represents the solutions $X = \{x_1, x_2, \ldots x_M\}$. In the mutation operation, a new candidate or mutant solution $\mathbf{m}_j^{s+1} = \left[m_{j,1}^{s+1}, m_{j,2}^{s+1}, \ldots, m_{j,d}^{s+1} \right]$ is generated for each individual $\mathbf{x}_j$ by adding the weighted difference between $\mathbf{x}_{ra_1}$ and $\mathbf{x}_{ra_2}$ to a third candidate solution $\mathbf{x}_{ra_3}$ as follows:

$$\mathbf{m}_j^{s+1} = \mathbf{x}_{ra_3}^s + \delta\left(\mathbf{x}_{ra_1}^s - \mathbf{x}_{ra_2}^s\right) \tag{8.8}$$

where $ra_1, ra_2, ra_3 \in \{1, 2, \ldots, M\}$ are subject to $ra_1 \neq ra_2 \neq ra_3 \neq j$ and denote the index of a randomly chosen solution. The parameter $\delta \in [0, 1]$ is the differential weight $(\mathbf{x}_{r_1}^s - \mathbf{x}_{r_2}^s)$, this magnitude is used to control the differential variation.

Additionally, in the crossover operation, DE generates a trial solution $\mathbf{u}_j^{s+1} = \left[u_{j,1}^{s+1}, u_{j,2}^{s+1}, \ldots, u_{j,d}^{s+1}\right]$ corresponding to all j members in the population. Here, $u_{j,n}^{s+1}$ is calculated as follows:

$$u_{j,n}^{s+1} = \begin{cases} m_{j,n}^{s+1} & \text{if (rand}(0, 1) \leq CR) \\ x_{j,n}^s & \text{if (rand}(0, 1) > CR) \text{ otherwise} \end{cases} \quad \text{for } n = 1, 2, \ldots, d \tag{8.9}$$

where, $n \in \{1, 2, \ldots, d\}$ is a dimension index chosen randomly, rand$(0, 1)$ is a random number within the interval $[0,1]$. Additionally, the parameter $CR \in [0, 1]$ represents a crossover rate which is used to regulate the probability of an element $u_{j,n}^{s+1}$ as a dimensional part of the mutant solution $\mathbf{m}_j^{s+1}$ $(m_{j,n}^{s+1})$ or a dimensional part from the candidate solution $\mathbf{x}_j^s$ $(x_{j,n}^s)$. This way to carried out the crossover is known as Binomial scheme [62].

Finally, in the selection process, each trail solution $\mathbf{u}_j^{s+1}$ is compared against its respective candidate solution $\mathbf{x}_j^s$ considering its fitness value. If the trial solution $\mathbf{u}_j^{s+1}$ yields a better fitness value than $\mathbf{x}_j^s$, then the value of the candidate solution for the next generation will be $\mathbf{u}_j^{s+1}$, otherwise, there will be no changes:

$$\mathbf{x}_j^{s+1} = \begin{cases} \mathbf{u}_j^{s+1} & \text{if } f\left(\mathbf{u}_j^{s+1}\right) > \text{if } f\left(\mathbf{x}_j^s\right) \\ \mathbf{x}_j^s & \text{otherwise} \end{cases} \tag{8.10}$$

For our experiments, a population of 25 individuals is considered. All individuals are subject to replacement and modification. The differential factor changes from 0.3 to 1 with a step of 0.1. The crossover probability CR is set to 0.5.

8.3 Methodology

In this section, the proposed methodology to segment vessel and optic disc is explained in detail. Even though the enhanced contrast produced by the Lateral Inhibition (LI) technique and the minimization of the cross-entropy function generated by the DE represent the main steps, the complete process includes other operations. The whole scheme is divided into three computing stages: pre-processing, processing and post-processing. Figure 8.1 shows graphically the effect of each process for both cases vessel (Fig. 8.1a) and optic disc (Fig. 8.1b) segmentation. For the sake of clarity, each process is divided into two parts: Vessel and optic disc segmentation.

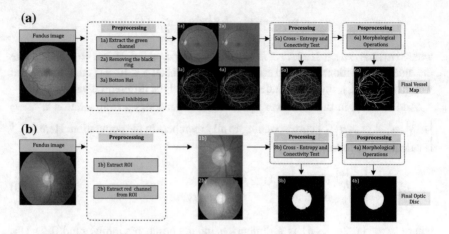

Fig. 8.1 The process diagram of the methodology. **a** Vessel segmentation **b** optic disc segmentation

8.3.1 Pre-processing

8.3.1.1 Vessel Segmentation

During the preprocessing, the information of the fundus image is modified so that the structure of vessels is highlighted. This process includes several procedures such as (1a) extracting the green channel information, (2a) removing the black ring, (3a) the bottom hat operation and (4a) lateral inhibition (LI).

The preprocessing stage aims to improve the image by reducing distortions or highlighting some features for further steps. Retinal vessel images usually show a weak contrast between the thin vessels and the background making them appear blurred [63]. Besides, thick vessels may appear thinner than they are. These effects are produced due to several factors such as the misalignment in the camera, poor focus, lighting problems, eye movement and low pupil dilation [64, 65]. Then, the preprocessing stage is fundamental to improve the image quality before the retinal vessel detection process.

In preprocessing, the fundus image is converted from RGB to grayscale by extracting the green channel. Green channel provides better contrast between the background and vessels [66, 67]. Additionally, the human eye maintains a better perception of the green channel than any of the other two channels. Therefore, the green channel is selected as the grayscale image (1a).

Then, the black background around the retina is removed (2a) from the grayscale image to provide uniformity in the image. A background with a high level of homogeneity is necessary to obtain better results when the bottom-hat filter is applied to the image [4]. Under such conditions, each black pixel is replaced with an intensity level equivalent to the average value within the retinal image.

After removing the black background around the retina, a bottom hat filter (3a) is applied. It is used to adjust the intensities improving the contrast of the image [65]. In its operation, the filter considers the application of a morphological operation of closing over the original image I_g. Then, the resulting image from this transformation $(I_g \cdot S)$ is subtracted from the original image [45]. Therefore, the bot-hat filter is defined as:

$$I_{bh} = (I_g \cdot S) - I_g, \tag{8.11}$$

where I_{bh} is the image after applying the filter. $(\cdot)$ is the closing operation while S corresponds to a standard linear structure element. The length of the structural element is 9×9 pixels. With the operation, a bottom-hat filter removes noisy areas and information not corresponding to the structure of vessels such as the optic disc and macula.

Finally, the resulting image is processed through the lateral inhibition (LI) method (described in Sect. 8.2.1). For our case, the set of LI coefficients is defined as follows [68]:

$$\delta_{ij} = \begin{bmatrix} -0.025 & -0.025 & -0.025 & -0.025 & -0.025 \\ -0.025 & -0.075 & -0.075 & -0.075 & -0.025 \\ -0.025 & -0.075 & 1 & -0.075 & -0.025 \\ -0.025 & -0.075 & -0.075 & -0.075 & -0.025 \\ -0.025 & -0.025 & -0.025 & -0.025 & -0.025 \end{bmatrix} \tag{8.12}$$

8.3.1.2 Optic Disc Segmentation

In the case of optic disc segmentation, the preprocessing stage has fewer steps in comparison to vessel segmentation; this is because the characteristics of the fundus image allow a direct analysis. The preprocessing scheme for optic disc segmentation includes the following processes: (1b) Region of interest (ROI) delimitation and (2b) the red channel extraction from ROI.

The optic disc (OD) images have elements that make the segmentation easier than the vessel. They maintain two main characteristics: first, the area of OD is larger than the vessels, and second, the OD is always in a specific region in the image.

8.3.2 Processing

8.3.2.1 Vessel Segmentation

The processing stage involves the use of the cross-entropy minimization. In this procedure, the cross-entropy is minimized to obtain the threshold that defines a pixel as a vessel or as part of the background. Essentially, the determination of the threshold value that minimize the cross-entropy is not a simple problem due to the non-linear nature of the resulting equations. Under such difficulties, the use of the DE algorithm is considered.

Cross-entropy minimization (Sect. 8.2.2) is applied to the retinal vessel image to segment it into two classes through a threshold value. A vessel or non-vessel classification depends on the threshold selection. To find the optimal threshold value, the DE algorithm is executed. Under the operation of DE (Sect. 8.2.3), a set of candidate solutions are given in every generation where every candidate solution represents a threshold value. The quality of each solution is evaluated through the cross entropy. According to the DE operators and the value of the objective function, new candidate solutions are generated along the process. As the method evolves, the quality of the solutions improves.

In general terms, the problem can be formulated as follows:

$$\min \quad f_{D_{CE}}(t)$$
$$\text{Subject to} : \mathbf{X} = \left\{ t \in \mathbb{R}^n \middle| 0 \le t_i \le 255, i = 1, 2, \ldots, n \right\} \tag{8.13}$$

where $f_{D_{CE}}(t)$ is the cross entropy function given by Eq. (8.8) and t represents a candidate solution.

8.3.2.2 Optic Disc Segmentation

Similar to the Vessel segmentation, the processing stage considers the use of the cross-entropy minimization. Under this operation, the cross-entropy is minimized through the operation of the DE method to obtain a specific threshold value. This value defines if a particular pixel belongs to the optic disc or the background.

8.3.3 Post-processing

Once the optimal threshold has been found, a post-processing stage is applied to eliminate noise or artifacts generated during the thresholding process. This stage is applied for both procedures Vessel and Optic disc segmentation.

The pre-processing is performed by using morphological operations such as closing and opening. These operations allow removing not connected structures considering as noise. Another important effect of such operations is to rebuild those elements considered as a part of vessel or the disc structures.

8.4   Experimental Results

This section presents the effectiveness of the proposed methodology when it is tested over three public data image sets. The DRIVE (Digital Retinal Images for Vessel Extraction) [69] and STARE (Structured Analysis of the Retina) [70] data sets for retinal vessel segmentation and DRISHTI-GS [71] for ODs. The results of our scheme are also compared with those obtained for other similar methods for vessel segmenting.

The DRIVE data set is composed of forty retinal images (565 × 584 pixels 8 bits per color channel) captured by a Canon CR5 nonmydriatic 3CCD camera with a 45° field of view. The data set is subdivided into the training and test group, each of twenty images. The images in the training group were manually segmented once, while the test case images were twice. The segmentations were performed by three different human observers previously trained by an ophthalmologist. The sets X and Y resulting from manual segmentation of the test case are used in this work as ground truth.

The STARE is a database with twenty images (605 × 700 pixels 8 bits per color channel) for blood vessel segmentation digitalized by a TopCon TVR-50 fundus camera with a 35° field of view. This dataset was manually segmented by two human observers. Where the first segmented the 10.4% as a vessel pixel while the second 14.9%. The results of this work use the segmentation of both observers as the ground truth.

The DRISHTI-GS dataset has 50 retinal fundus images extracted from glaucoma patients. Each image has been acquired considering a field of view of 30 degrees centered on the optic disc and size of 28,840 × 1944 pixels. The ground truth images were determined by four ophthalmologist experts with 3, 5, 9, and 20 years of clinical expertise.

In the segmentation processes, cross-entropy thresholding is applied to partition the image into two classes by determining a threshold value. To make an appropriate threshold selection, DE algorithm is used to minimize the cross-entropy between classes. A set of 25 candidate solutions are given in every generation and the maximum number of generations is 10. The experiments were executed 30 times to verify the consistency of the results.

The result of the segmentation process can be approached as a classification process where a pixel belongs to a vessel, disc structure or background. To evaluate the performance of correct classification, three measurements are used as performance indexes: Sensitivity (Se), Specificity (Sp), and Accuracy (Acc). Se reflects the capability of the algorithm to detect vessel or disc pixels, Sp is the ability to detect the non-vessel or non-disc pixels, and Acc evaluates the confidence of the algorithm. These measurements are defined as follows:

$$Sensitivity\,(Se) = \frac{T_P}{T_P + F_N} \qquad\qquad (8.14)$$

$$Specificity\,(Sp) = \frac{T_N}{T_N + F_P} \qquad\qquad (8.15)$$

$$Accuracy\,(Acc) = \frac{T_P + T_N}{T_P + T_N + F_P + F_N} \qquad\qquad (8.16)$$

where T_P (True Positive) indicates the number of pixels classified as vessel or disc elements in both, the segmented image and the ground truth. T_N (True Negative) corresponds to the number of pixels considered as non-vessel or mom-disc elements in both, ground truth and segmented image. In the other hand, F_P (False Positive) represents the among of pixels considered as vessel or disc elements in the segmented image, but non-vessel or non-disc in the ground truth. F_N (False Negative) exhibits the number of pixels classified as non-vessel or non-disc elements in the segmented image, when they correspond to vessel or disc elements in the ground truth.

In addition to these indexes, the F-score is also considered to evaluate the Optic disc (OD) segmentation results. F-score is proposed by the dataset authors to measure accurately the segmentation performance of a specific method. The F-score corresponds to a value between 0 and 1. The value of 0 represents the worst performance while the value of 1 corresponds to the best one. The F-score is defined by the following formulation:

$$F - score = \frac{(2 * Pr * Re)}{Pr + Re} \qquad\qquad (8.17)$$

where Pr and Re are the Precision and Recall measurements that are calculated by the following expressions:

$$Precision\,(Pr) = \frac{T_P}{T_P + F_P} \qquad\qquad (8.18)$$

$$Recall\,(Re) = \frac{T_P}{T_P + F_N} \qquad\qquad (8.19)$$

Table 8.1 shows the results of the proposed methodology for both observers in DRIVE database, Table 8.2 presents the results of STARE database and Table 8.3 presents the results of DRISHTI-GS dataset from twenty test cases. The reported results are the mean values of the 30 independent executions in order to avoid random effects. According to the results exhibited in both tables, the proposed method obtains competitive values in terms of its respective performance indexes. It is also remarkable that from all indexes the proposed approach presents the best values in the specificity and accuracy indexes.

Figure 8.2 illustrates the segmentation results of the proposed methodology over a representative set of six images in comparison with the ground truth images. The first

Table 8.1 Results of the proposed algorithm in DRIVE database images

Image	Drive observer 1			Drive observer 2		
	Se	Sp	Acc	Se	Sp	Acc
1	0.8316	0.9730	0.9621	0.8359	0.9753	0.9645
2	0.8918	0.9606	0.9555	0.9235	0.9649	0.9618
3	0.8360	0.9564	0.9479	0.8306	0.9675	0.9578
4	0.8793	0.9613	0.9561	0.8952	0.9671	0.9625
5	0.8714	0.9596	0.9540	0.8770	0.9733	0.9671
6	0.8378	0.9574	0.9492	0.8588	0.9634	0.9562
7	0.8516	0.9594	0.9527	0.8004	0.9771	0.9660
8	0.8121	0.9596	0.9509	0.7442	0.9760	0.9623
9	0.8532	0.9655	0.9591	0.8766	0.9671	0.9620
10	0.8250	0.9684	0.9592	0.8077	0.9785	0.9676
11	0.8677	0.9632	0.9572	0.8922	0.9716	0.9666
12	0.8003	0.9680	0.9561	0.8052	0.9748	0.9628
13	0.8457	0.9563	0.9489	0.8789	0.9551	0.9499
14	0.8030	0.9708	0.9596	0.8018	0.9773	0.9656
15	0.7713	0.9758	0.9629	0.7885	0.9738	0.9620
16	0.8812	0.9671	0.9613	0.8722	0.9720	0.9652
17	0.8540	0.9676	0.9604	0.8246	0.9768	0.9672
18	0.8019	0.9750	0.9629	0.8857	0.9677	0.9620
19	0.8510	0.9792	0.9696	0.8766	0.9637	0.9572
20	0.7704	0.9737	0.9608	0.8531	0.9589	0.9522
Mean	0.8368	0.9659	0.9573	0.8464	0.9701	0.9619

four are from DRIVE database and the other two from STARE database. Figure 8.3 shows the optic disc segmentation results over four images from the DRISHTI-GS. In Tables 8.4 and 8.5, numerical comparisons between the proposed methodology results and other up-to-date methods are exhibited. In the case of optic disc segmentation, the results are taken from [29]. Tables 8.4 and 8.5 shows that the accuracy of the proposed method in both segmentation processes is superior. Our method also demonstrates that it reaches a good performance whiteout requiring a training stage or a machine learning technique. Figure 8.4 exhibits the sensitivity evaluation for the ROC curve and AUC values for four images considering disc segmentation.

8.5 Conclusions

In this chapter, an accurate methodology for retinal vessel and optic disc segmentation has been presented. The proposed method combines two different techniques the

Table 8.2 Results of the proposed algorithm in STARE database

Image	Stare observer 1			Stare observer 2		
	Se	Sp	Acc	Se	Sp	Acc
1	0.8488	0.9575	0.9524	0.8426	0.9439	0.9392
2	0.7595	0.9528	0.9475	0.6646	0.9615	0.9534
3	0.7998	0.9770	0.9686	0.6821	0.9735	0.9596
4	0.8224	0.9466	0.9432	0.7019	0.9452	0.9386
5	0.8944	0.9574	0.9539	0.8936	0.9418	0.9391
6	0.8368	0.9453	0.9433	0.9392	0.9019	0.9026
7	0.7782	0.9669	0.9550	0.9322	0.9242	0.9247
8	0.6877	0.9624	0.9468	0.8830	0.9077	0.9063
9	0.8737	0.9753	0.9688	0.9606	0.9343	0.9360
10	0.7948	0.9581	0.9497	0.9541	0.8904	0.8936
11	0.8439	0.9793	0.9710	0.9641	0.9411	0.9425
12	0.8860	0.9761	0.9705	0.9852	0.9415	0.9442
13	0.8765	0.9611	0.9561	0.9814	0.9124	0.9165
14	0.8814	0.9594	0.9548	0.9560	0.9193	0.9215
15	0.8459	0.9566	0.9507	0.9562	0.9217	0.9235
16	0.8212	0.9293	0.9248	0.9204	0.8810	0.8827
17	0.9140	0.9617	0.9589	0.9761	0.9213	0.9245
18	0.9050	0.9767	0.9745	0.9503	0.9666	0.9661
19	0.8142	0.9819	0.9766	0.8931	0.9641	0.9619
20	0.7777	0.9567	0.9510	0.8753	0.9259	0.9243
Mean	0.8331	0.9619	0.9559	0.8956	0.9310	0.9300

Lateral Inhibition (LI) and the Differential Evolution (DE). The LI scheme produces a new image with an enhanced contrast between the background and retinal vessels. Then, the DE algorithm is used to obtain the appropriate threshold values through the minimization of the cross-entropy function from the enhanced image. Finally, morphological operations are applied to the final image for eliminating noise and artifacts in the output image.

To evaluate the performance of the proposed approach, several experiments over images extracted from STARE and DRIVE databases for retinal vessel and DRISHTI-GS dataset for optic disc segmentation have been conducted. Simulation results demonstrate a high performance of the proposed scheme in comparison with similar methods reported in the literature.

Table 8.3 Results of the proposed algorithm in DRISHTI-GS database

Image	Se	Sp	Acc	F-score
1	0.9775	0.9996	0.9986	0.9844
2	0.9991	0.9989	0.9989	0.9816
3	0.9831	0.9995	0.9991	0.9810
4	0.9981	0.9990	0.9990	0.9810
5	0.9905	0.9991	0.9989	0.9809
6	0.9992	0.9983	0.9983	0.9761
7	0.9901	0.9987	0.9985	0.9754
8	0.9822	0.9990	0.9985	0.9736
9	0.9839	0.9992	0.9989	0.9724
10	0.9996	0.9984	0.9984	0.9723
11	0.9977	0.9983	0.9983	0.9703
12	0.9822	0.9987	0.9982	0.9699
13	0.9977	0.9981	0.9981	0.9696
14	0.9759	0.9989	0.9982	0.9694
15	0.9842	0.9985	0.9980	0.9685
16	0.9308	0.9999	0.9981	0.9627
17	0.9908	0.9983	0.9981	0.9625
18	0.9976	0.9978	0.9978	0.9619
19	0.9989	0.9970	0.9970	0.9615
20	0.9374	0.9996	0.9979	0.9608
Min	0.9308	0.9880	0.9883	0.8319
Max	0.9998	0.9999	0.9991	0.9844
Mean	0.9889	0.9970	0.9968	0.9493
Std	0.0151	0.0026	0.0024	0.0340

Image	Observer 1	Observer 2	Segmentation	Database

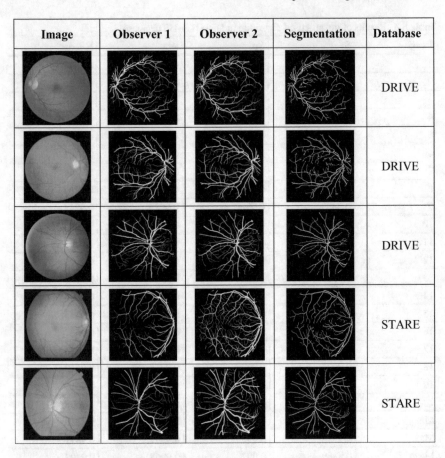

Fig. 8.2 Segmented images by proposed methodology

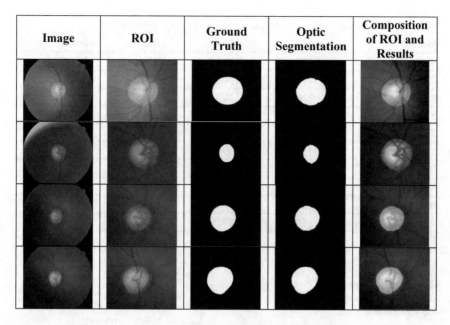

Image	ROI	Ground Truth	Optic Segmentation	Composition of ROI and Results

Fig. 8.3 Segmented images of the optic disc by the proposed methodology

Table 8.4 Comparison results with related works for vessel segmentation

Methods	Drive			Stare		
	Se	Sp	Acc	Se	Sp	Acc
Jiang et al. [72]	–	–	0.9212	–	–	–
Zhang et al. [18]	0.7120	0.9724	0.9382	0.7171	0.9753	0.9483
Staal et al. [69]	0.7194	0.9773	0.9442	0.6970	0.9810	0.9516
Qian et al. [1]	0.7354	0.9789	0.9477	0.7187	0.9767	0.9509
Câmara Neto et al. [73]	0.7942	0.9632	–	0.7695	0.9537	0.8616
Rezaee et al. [74]	0.7189	0.9793	0.9463	0.7202	0.9741	0.9521
Zhao [1]	0.7354	0.9789	0.9477	0.7187	0.9767	0.9509
Rodrigues et al. [75]	0.7165	0.9801	0.9465	–	–	–
Marin et al. [76]	0.7067	0.9801	0.9452	0.6944	0.9819	0.9526
Proposed method	**0.8464**	**0.9701**	**0.9619**	**0.8331**	**0.9619**	**0.9559**

Table 8.5 Comparison results with related works for optic disc segmentation

Methods	Se	Sp	Acc	F-score
Wong et al. [30]	N/A	N/A	N/A	91.10
Cheng et al. [31]	N/A	N/A	N/A	92.10
Sedai et al. [32]	N/A	N/A	N/A	95.00
Zilly et al. [28]	N/A	N/A	N/A	97.30
Stapor et al. [33]	84.98	99.64	N/A	90.20
Seo et al. [34]	61.03	**99.87**	N/A	N/A
Lupascu et al. [11]	68.48	99.69	N/A	N/A
Kande et al. [10]	88.08	98.78	N/A	N/A
Bharkad et al. [35]	74.60	99.61	N/A	N/A
Almotiri et al. [29]	93.13	97.09	N/A	N/A
Proposed method	**98.89**	99.70	**99.68**	**94.93**

Fig. 8.4 ROC curve and AUC values of four optic disc segmentation

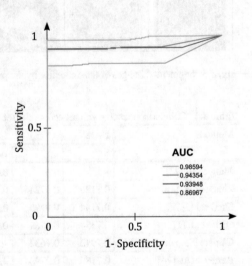

References

1. Zhao, Y.Q., Wang, X.H., Wang, X.F., Shih, F.Y.: Retinal vessels segmentation based on level set and region growing. Pattern Recognit. **47**(7), 2437–2446 (2014)
2. Stanton, A.V., et al.: Vascular network changes in the retina with age and hypertension. J. Hypertens. **13**(12 Pt 2), 1724–1728 (1995)
3. Skovborg, F., Nielsen, A.V., Lauritzen, E., Hartkopp, O.: Diameters of the retinal vessels in diabetic and normal subjects. Diabetes **18**(5), 292–298 (1969)
4. Martinez-Perez, M.E., Hughes, A.D., Thom, S.A., Bharath, A.A., Parker, K.H.: Segmentation of blood vessels from red-free and fluorescein retinal images. Med. Image Anal. **11**(1), 47–61 (2007)
5. Lázár, I., Hajdu, A.: Segmentation of retinal vessels by means of directional response vector similarity and region growing. Comput. Biol. Med. **66**, 209–221 (2015)

6. Fraz, M.M., et al.: Blood vessel segmentation methodologies in retinal images—A survey. Comput. Methods Programs Biomed. **108**(1), 407–433 (2012)

7. Kirbas, C., Quek, F.: A review of vessel extraction techniques and algorithms (2003)

8. Franklin, S.W., Rajan, S.E.: Retinal vessel segmentation employing ANN technique by Gabor and moment invariants-based features. Appl. Soft Comput. **22**, 94–100 (2014)

9. Wang, S., Yin, Y., Cao, G., Wei, B., Zheng, Y., Yang, G.: Hierarchical retinal blood vessel segmentation based on feature and ensemble learning. Neurocomputing **149**, 708–717 (2015)

10. Kande, G.B., Subbaiah, P.V., Savithri, T.S.: Unsupervised fuzzy based vessel segmentation in pathological digital fundus images. J. Med. Syst. **34**(5), 849–858 (2010)

11. Lupaşcu, C.A., Tegolo, D.: Automatic Unsupervised Segmentation of Retinal Vessels Using Self-Organizing Maps and K-Means Clustering, pp. 263–274. Springer, Berlin, Heidelberg (2011)

12. Yin, Y., Adel, M., Bourennane, S.: Automatic segmentation and measurement of vasculature in retinal fundus images using probabilistic formulation. Comput. Math. (2013)

13. Zhang, J., Li, H., Nie, Q., Cheng, L.: A retinal vessel boundary tracking method based on Bayesian theory and multi-scale line detection. Comput. Med. Imaging Graph. **38**(6), 517–525 (2014)

14. Zhang, Y., Hsu, W., Lee, M.L.: Detection of retinal blood vessels based on nonlinear projections. J. Signal Process. Syst. **55**(1–3), 103–112 (2009)

15. Khdhair, N., Abbadi, E., Hamood, E., Saadi, A.: Blood vessels extraction using mathematical morphology. J. Comput. Sci. Publ. Online **9**(910), 1389–1395 (2013)

16. Hassan G, El-Bendary N, Hassanien AE, Fahmy A, Snasel V.: Retinal blood vessel segmentation approach based on mathematical morphology. Proc. Comput. Sci. **65**, 612–622 (2015)

17. Kumar, D., Pramanik, A., Kar, S.S., Maity, S.P.: Retinal blood vessel segmentation using matched filter and Laplacian of Gaussian. In: 2016 International Conference on Signal Processing and Communications (SPCOM), 2016, pp. 1–5

18. Zhang, B., Zhang, L., Zhang, L., Karray, F.: Retinal vessel extraction by matched filter with first-order derivative of Gaussian. Comput. Biol. Med. **40**(4), 438–445 (2010)

19. Ben Abdallah, M., et al.: Automatic extraction of blood vessels in the retinal vascular tree using multiscale medialness. Int. J. Biomed. Imaging **2015** (2015)

20. Ryu, H., Moon, H., Browatzki, B., Wallraven, C.: Retinal vessel detection using deep learning: a novel Directnet architecture. Korean J. Vis. Sci. **20**(2), 151–159 (2018)

21. Oliveira, A., Pereira, S., Silva, C.A.: Retinal vessel segmentation based on fully convolutional neural networks. Expert Syst. Appl. **112**, 229–242 (2018)

22. Kaur, J., Mittal, D.: A generalized method for the detection of vascular structure in pathological retinal images. Biocybern. Biomed. Eng. **37**(1), 184–200 (2017)

23. Jiang, Z., Yepez, J., An, S., Ko, S.: Fast, accurate and robust retinal vessel segmentation system. Biocybern. Biomed. Eng. (2017)

24. Khakzar, M., Pourghassem, H.: A retinal image authentication framework based on a graph-based representation algorithm in a two-stage matching structure. Biocybern. Biomed. Eng. **37**(4), 742–759 (2017)

25. GeethaRamani, R., Balasubramanian, L.: Retinal blood vessel segmentation employing image processing and data mining techniques for computerized retinal image analysis. Biocybern. Biomed. Eng. **36**(1), 102–118 (2016)

26. Aslani, S., Sarnel, H.: A new supervised retinal vessel segmentation method based on robust hybrid features. Biomed. Signal Process. Control **30**, 1–12 (2016)

27. Christodoulidis, A., Hurtut, T., Tahar, H.B., Cheriet, F.: A multi-scale tensor voting approach for small retinal vessel segmentation in high resolution fundus images. Comput. Med. Imaging Graph. **52**, 28–43 (2016)

28. Zilly, J., Buhmann, J.M., Mahapatra, D.: Glaucoma detection using entropy sampling and ensemble learning for automatic optic cup and disc segmentation. Comput. Med. Imaging Graph. **55**, 28–41 (2017)

29. Almotiri, J., Elleithy, K., Elleithy, A.: A multi-anatomical retinal structure segmentation system for automatic eye screening using morphological adaptive fuzzy thresholding. IEEE J. Transl. Eng. Heal. Med. **6**, 1–23 (2018)
30. Wong, D.W.K., et al.: Level-set based automatic cup-to-disc ratio determination using retinal fundus images in ARGALI. In: 2008 30th Annual International Conference of the IEEE Engineering in Medicine and Biology Society, 2008, pp. 2266–2269
31. Cheng, J., et al.: Superpixel classification based optic disc and optic cup segmentation for glaucoma screening. IEEE Trans. Med. Imaging **32**(6), 1019–1032 (2013)
32. Sedai, S., Roy, P.K., Mahapatra, D., Garnavi, R.: Segmentation of optic disc and optic cup in retinal fundus images using shape regression. In: 2016 38th Annual International Conference of the IEEE Engineering in Medicine and Biology Society (EMBC), 2016, pp. 3260–3264
33. Stąpor, K., Świtonski, A., Chrastek, R., Michelson, G.: Segmentation of Fundus Eye Images Using Methods of Mathematical Morphology for Glaucoma Diagnosis, pp. 41–48. Springer, Berlin, Heidelberg, (2004)
34. Seo, J.M., Kim, K.K., Kim, J.H., Park, K.S., Chung, H.: Measurement of ocular torsion using digital fundus image. In: The 26th Annual International Conference of the IEEE Engineering in Medicine and Biology Society, vol. 3, pp. 1711–1713
35. Bharkad, S.: Automatic segmentation of optic disk in retinal images. Biomed. Signal Process. Control **31**, 483–498 (2017)
36. Cuevas, E., Sención-Echauri, F., Zaldivar, D., Pérez-Cisneros, M.: Multi-circle detection on images using artificial bee colony (ABC) optimization. Soft. Comput. **16**(2), 281–296 (2012)
37. Oliva, D., Cuevas, E.: A Medical Application: Blood Cell Segmentation by Circle Detection, pp. 135–157. Springer, Cham (2017)
38. Oliva, D., Hinojosa, S., Cuevas, E., Pajares, G., Avalos, O., Gálvez, J.: Cross entropy based thresholding for magnetic resonance brain images using crow search algorithm. Expert Syst. Appl. **79**, 164–180 (2017)
39. González, A., Cuevas, E., Fausto, F., Valdivia, A., Rojas, R.: A template matching approach based on the behavior of swarms of locust. Appl. Intell., pp. 1–12 (May 2017)
40. Valdivia-Gonzalez, A., Zaldívar, D., Fausto, F., Camarena, O., Cuevas, E., Perez-Cisneros, M.: A states of matter search-based approach for solving the problem of intelligent power allocation in plug-in hybrid electric vehicles. Energies **10**(1), 92 (2017)
41. Yang, Y., Wang, Z., Yang, B., Jing, Z., Kang, Y.: Multiobjective optimization for fixture locating layout of sheet metal part using SVR and NSGA-II. Math. Probl. Eng. **2017**, 1–10 (2017)
42. Zhang, H., Dai, Z., Zhang, W., Zhang, S., Wang, Y., Liu, R.: A new energy-aware flexible job shop scheduling method using modified biogeography-based optimization. Math. Probl. Eng. **2017**, 1–12 (2017)
43. Pang, C., Huang, S., Zhao, Y., Wei, D., Liu, J.: Sensor network disposition facing the task of multisensor cross cueing. Math. Probl. Eng. **2017**, 1–8 (2017)
44. Kóczy, L.T., Földesi, P., Tüű-Szabó, B.: An effective discrete bacterial memetic evolutionary algorithm for the traveling salesman problem. Int. J. Intell. Syst. **32**(8), 862–876 (2017)
45. Bai, X., Zhou, F.: Multi structuring element top-hat transform to detect linear features. In: IEEE 10th International Conference on Signal Processing Proceedings, 2010, pp. 877–880
46. Céspedes-Mota, A., et al.: Optimization of the distribution and localization of wireless sensor networks based on differential evolution approach. Math. Probl. Eng. **2016**, 1–12 (2016)
47. Lai, L., Ji, Y.-D., Zhong, S.-C., Zhang, L.: Sequential parameter identification of fractional-order duffing system based on differential evolution algorithm. Math. Probl. Eng. **2017**, 1–13 (2017)
48. Bhattacharyya, S., Konar, A., Tibarewala, D.N.: A differential evolution based energy trajectory planner for artificial limb control using motor imagery EEG signal. Biomed. Signal Process. Control **11**(1), 107–113 (2014)
49. Elsayed, S., Sarker, R.: Differential evolution framework for big data optimization. Memetic Comput. **8**(1), 17–33 (2016)
50. Hartline, H.K.: The response of single optic nerve fibers of the vertebrate eye to illumination of the Retina. Am. J. Physiol. Content **121**(2), 400–415 (1938)

51. Li, B., Li, Y., Cao, H., Salimi, H.: Image enhancement via lateral inhibition: an analysis under illumination changes. Optik (Stuttg) **127**(12), 5078–5083 (2016)
52. Fang, Z., Dawei, Z., Ke, Z.: Image pre-processing algorithm based on lateral inhibition. In: 2007 8th International Conference on Electronic Measurement and Instruments, 2007, pp. 2-701–2-705
53. Cormack, R.H., Coren, S., Girgus, J.S.: Seeing is deceiving: the psychology of visual illusions. Am. J. Psychol. **92**(3), 557 (1979)
54. Liu, F., Duan, H., Deng, Y.: A chaotic quantum-behaved particle swarm optimization based on lateral inhibition for image matching. Opt. - Int. J. Light Electron Opt. **123**(21), 1955–1960 (2012)
55. Kullback, S.: Information Theory and Statistics. Dover Publications (1968)
56. Li, C.H., Lee, C.K.: Minimum cross entropy thresholding. Pattern Recognit. **26**(4), 617–625 (1993)
57. Cuevas, E.: Block-matching algorithm based on harmony search optimization for motion estimation. Appl. Intell. **39**(1), 165–183 ((2013))
58. Díaz-Cortés, M.-A., Ortega-Sánchez, N., Hinojosa, S., Cuevas, E., Rojas, R., Demin, A.: A multi-level thresholding method for breast thermograms analysis using Dragonfly algorithm. Infrared Phys. Technol. **93**, 346–361 (2018)
59. Díaz, P., Pérez-Cisneros, M., Cuevas, E., Hinojosa, S., Zaldivar, D.: An im-proved crow search algorithm applied to energy problems. Energies **11**(3), 571 (2018)
60. Storn, R., K.P.-J. of Global Optimization, and Undefined 1997, "Differential Evolution—A Simple and Efficient Heuristic for Global Optimization Over Continuous Spaces. Springer, Berlin
61. Schmitt, L.M.: Theory of genetic algorithms. Theor. Comput. Sci. **259**(1), 1–61 (2001)
62. Yang, X.-S.: Nature-inspired optimization algorithms. In: Nature-Inspired Optimization Algorithms (2014)
63. Rahebi, J., Hardalaç, F.: Retinal blood vessel segmentation with neural network by using gray-level co-occurrence matrix-based features. J. Med. Syst. **38**(8), 85 (2014)
64. Zheng, Y., Kwong, M.T., Maccormick, I.J.C., Beare, N.A.V., Harding, S.P.: A comprehensive texture segmentation framework for segmentation of capillary non-perfusion regions in fundus fluorescein angiograms. PLoS One **9**(4) (2014)
65. Bai, X., Zhou, F., Xue, B.: Image enhancement using multi scale image features extracted by top-hat transform. Opt. Laser Technol. **44**(2), 328–336 (2012)
66. Salazar-Gonzalez, A., Kaba, D., Li, Y., Liu, X.: Segmentation of the blood vessels and optic disk in retinal images. IEEE J. Biomed. Heal. Informatics **18**(6), 1874–1886 (2014)
67. Soares, J.V.B., Leandro, J.J.G., Cesar, R.M., Jelinek, H.F., Cree, M.J.: Retinal vessel segmentation using the 2-D Gabor wavelet and supervised classification. IEEE Trans. Med. Imaging **25**(9), 1214–1222 (2006)
68. Wang, X., Duan, H., Luo, D.: Cauchy biogeography-based optimization based on lateral inhibition for image matching. Optik (Stuttg) **124**(22), 5447–5453 (2013)
69. Staal, J., Abramoff, M.D., Niemeijer, M., Viergever, M.A., van Ginneken, B.: Ridge-based vessel segmentation in color images of the Retina. IEEE Trans. Med. Imaging **23**(4), 501–509 (2004)
70. Hoover, A.D., Kouznetsova, V., Goldbaum, M.: Locating blood vessels in retinal images by piecewise threshold probing of a matched filter response. IEEE Trans. Med. Imaging **19**(3), 203–210 (2000)
71. Sivaswamy, J., Krishnadas, S.R., Datt Joshi, G., Jain, M., Syed Tabish, A.U.: Drishti-GS: Retinal image dataset for optic nerve head (ONH) segmentation, pp. 53–56 (2014)
72. Jiang, X., Mojon, D.: Adaptive local thresholding by verification-based multithreshold probing with application to vessel detection in retinal images. IEEE Trans. Pattern Anal. Mach. Intell. **25**(1), 131–137 (2003)
73. Neto, L.C., Ramalho, G.L.B., Neto, J.R., Veras, R.M.S., Medeiros, F.N.S.: An unsupervised coarse-to-fine algorithm for blood vessel segmentation in fundus images. Expert Syst. Appl. **78**(C), 182–192 (2017)

74. Rezaee, K., Haddadnia, J., Tashk, A.: Optimized clinical segmentation of retinal blood vessels by using combination of adaptive filtering, fuzzy entropy and skeletonization. Appl. Soft Comput. J. **52**, 937–951 (2017)
75. Rodrigues, L.C., Marengoni, M.: Segmentation of optic disc and blood vessels in retinal images using wavelets, mathematical morphology and Hessian-based multi-scale filtering. Biomed. Signal Process. Control **36**, 39–49 (2017)
76. Marín, D., Aquino, A., Gegundez-Arias, M.E., Bravo, J.M.: A new supervised method for blood vessel segmentation in retinal images by using gray-level and moment invariants-based features. IEEE Trans. Med. Imaging **30**(1), 146–158 (2011)

Chapter 9
Detection of White Blood Cells with Metaheuristic Computation

Abstract This chapter illustrates the use of metaheuristic computation schemes for the automatic detection of white blood cells embedded into complicated and cluttered smear images. The approach considers the identification problem as the process of detection of multi-ellipse shapes. The scheme uses the Differential Evolution (DE) method, which is easy to use, maintains a quite simple computation scheme presenting acceptable convergence properties. The approach considers the use of five edge points as agents which represent the candidate ellipses in the edge image of the smear. A cost function assesses if such candidate ellipses are present in the actual edge image. With the values of the cost function, the set of agents is modified by using the DE algorithm so that they can approximate the white blood cells contained in the edge-only map of the image.

9.1 Introduction

Medical computer vision is important in diagnosis with the development of medical imaging and computer technology. X-ray radiography, CT and MRI produce a significant number of medical images. They give essential information for effective and correct diagnosis based on advanced computer vision methods [1, 2].

On the other hand, White Blood Cells (WBC), also known as leukocytes, play an essential role in the analysis of various diseases. Although image processing methods have favorably contributed to producing new techniques for cell analysis, they have provided more accurate and reliable systems for disease diagnosis. However, the large variability on cell shape, size, edge and position, obscures the data extraction process. Furthermore, the difference between cell borders and the image's background may change due to variable lighting circumstances during the capturing process.

Several efforts have been achieved in the area of blood cell detection. In [3] a system based on edge support vectors is introduced to recognize WBC. In this approach, the intensity of each pixel is utilized to create feature vectors, whereas a Support Vector Machine (SVM) is employed to classify or segment. By considering another scheme, in [4], Wu et al. introduced an iterative Otsu method based on the

© Springer Nature Switzerland AG 2021
E. Cuevas et al., *Metaheuristic Computation: A Performance Perspective*,
Intelligent Systems Reference Library 195,
https://doi.org/10.1007/978-3-030-58100-8_9

circular histogram for leukocyte segmentation. According to this approach, the smear images are converted in the Hue-Saturation-Intensity (HSI) space, considering that the Hue plane includes most WBC data. One of the most modern advancements in white blood cell detection analysis is the method proposed by Wang [5] that is based on the fuzzy cellular neural network (FCNN). Although this technique has demonstrated success in recognizing only one leukocyte in the image, it has not been examined over images holding different white cells. Furthermore, its performance generally worsens when the iteration number is not defined correctly, yielding a challenging problem itself with no clear evidence on how to make the best choice.

Since WBC can be represented with an ellipsoid shape, image processing methods for identifying ellipses may be adopted to detect them. Ellipse recognition in actual images is an open research question for a long time ago. Several techniques have been suggested, which traditionally can be divided into three classes: Symmetry-based, Hough transform-based (HT) and Random sampling.

In symmetry-based methods [6, 7], the ellipse geometry is considered for its detection. The most basic components used for ellipse geometry are the ellipse center and axis. Using such parameters and the edges from the image, the ellipse parameters can be detected. The most popular method of this category is the Hough Transform (HT) [8]. HT operates by describing the geometric shape by a set of parameters. Then, a histogram is produced through the accumulation and quantization of the parameter values. The peaks in the histograms indicate where ellipses may be located. The intervals of the histogram bins directly affect the accuracy of the results and the computational effort since the parameters are quantized into discrete histograms. Therefore, for small quantization of the parameter space, the method yields more reliable results, while experiencing large memory requirements and expensive computation. To solve such this problem, some researchers have suggested other ellipse detectors following random sampling. In random sampling-based methods [9, 10], a histogram bin describes a candidate shape rather than a set of quantized parameters, as in the HT. Nevertheless, similar to the HT, random sampling methods operate through an aggregation process for the bins. The bin with the essential score describes the best estimate of an actual ellipse in the original image. McLaughlin's study [11] confirms that a random sampling-based method provides improvements in accuracy and computational complexity, as well as an elimination of the number of false positives (non-existent ellipses), when it is compared to the original HT.

The problem of ellipse detection has also been handled through optimization methods as an option to conventional techniques. In general, they have shown better results than those based on the HT and random sampling concerning the accuracy, speed and robustness [12]. Such methods have originated different robust ellipse detectors using various optimization schemes such as Genetic algorithms (GA) [13, 14] and Particle Swarm Optimization (PSO) [15].

Although detection schemes based on optimization techniques offer distinct advantages compared to conventional methods, they have barely been implemented to WBC detection. One exemption is the study presented by Karkavitsas and Rangoussi [16] that proposes the use of GA to solve the WBC detection problem. However, since the evaluation function considers the number of pixels inside a circle with a fixed radius, the method is prone to produce misdetections, particularly for images that contained overlapped or irregular WBC.

In this chapter, the WBC detection problem is faced as an optimization process. Under such conditions, the differential evolution algorithm is employed to produce an ellipsoidal approximation. Differential Evolution (DE), proposed by Storn and Price [17], refers to a metaheuristic method which is considered to obtain the optimal value for complex continuous nonlinear functions. As a population-based scheme, DE employs simple crossover and mutation schemes to produce new solutions and uses a greedy competition scheme to select whether the new candidate or its parent will be maintained in the next generation. Due to its characteristics, the DE algorithm has attracted much attention, being used in a wide range of successful applications in the literature [18–22].

9.2 Differential Evolution

Metaheuristic schemes have been demonstrated its interesting properties in several domains [8]. DE is a metaheuristic method that employs a mutation and a crossover operator as mechanism to provide the exchange of information among different agents.

There are several mutation types to define the operation of DE. The version of DE method considered in this chapter is called the rand-to-best/1/bin or "DE1" [23]. DE schemes start by initializing a population of N_p and D-dimensional vectors where parameter values are randomly distributed between the pre-specified lower initial parameter bound $x_{j,\text{low}}$ and the upper initial parameter bound $x_{j,\text{high}}$ as follows:

$$x_{j,i,t} = x_{j,\text{low}} + \text{rand}(0, 1) \cdot (x_{j,\text{high}} - x_{j,\text{low}});$$
$$j = 1, 2, \ldots, D; \quad i = 1, 2, \ldots, N_p; \quad t = 0. \tag{9.1}$$

The subscript t represents the iteration number, while j and i correspond to the parameter and agent indexes, respectively. Therefore, $x_{j,i,t}$ symbolizes the jth decision variable of the ith solution in iteration t. To produce a trial solution, the DE method first mutates the best solution $\mathbf{x}_{best,t}$ from the current population by combining the scaled difference of two different solutions from the population.

$$\mathbf{v}_{i,t} = \mathbf{x}_{best,t} + F \cdot (\mathbf{x}_{r_1,t} - \mathbf{x}_{r_2,t});$$
$$r_1, r_2 \in \{1, 2, \ldots, N_p\} \tag{9.2}$$

where $\mathbf{v}_{i,t}$ represents the mutant result vector. The indices r_1 and r_2 represent randomly selected agents with the constraint that they have no relation to the particle index i whatsoever (i.e., $r_1 \neq r_2 \neq i$). The mutation scale factor F represents a positive value, commonly less than one. Figure 9.1 exemplifies the vector-operation process specified by Eq. (9.2).

Fig. 9.1 Two-dimensional
illustration of the process for
generating v considering a
certain agent distribution

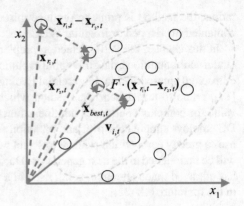

To enhance the diversity of the solution vector, the crossover operator is used between the mutant vector $\mathbf{v}_{i,t}$ and the original solutions $\mathbf{x}_{i,t}$. The product of this operation is the trial vector $\mathbf{u}_{i,t}$ that is calculated element to element as follows:

$$u_{j,i,t} = \begin{cases} v_{j,i,t}, & \text{if rand}(0,1) \le CR \text{ or } j = j_{\text{rand}}, \\ x_{j,i,t}, & \text{otherwise.} \end{cases} \tag{9.3}$$

where $j_{\text{rand}} \in \{1, 2, \ldots, D\}$. The crossover factor $(0.0 \le CR \le 1.0)$ regulates the number of decision variables that the mutant vector contributes to the final trial vector. Additionally, the trial vector always inherits the mutant vector parameter according to the randomly chosen index j_{rand}, assuring that the trial vector differs by at least one parameter from the vector to which it is compared $(\mathbf{x}_{i,t})$.

As a final step, a greedy selection scheme is considered to obtain better solutions. Under this step, if the cost value of $\mathbf{u}_{i,t}$ is less or equal than the cost value of $\mathbf{x}_{i,t}$, then $\mathbf{u}_{i,t}$ replaces $\mathbf{x}_{i,t}$ in the next iteration. Otherwise, $\mathbf{x}_{i,t}$ remains without a change in the population:

$$\mathbf{x}_{i,t+1} = \begin{cases} \mathbf{u}_{i,t}, & \text{if } f(\mathbf{u}_{i,t}) \le f(\mathbf{x}_{i,t}), \\ \mathbf{x}_{i,t}, & \text{otherwise.} \end{cases} \tag{9.4}$$

where, $f()$ represents the cost function. This process is repeated until a termination criterion has been reached.

9.3   Ellipse Detection Under an Optimization Perspective

To identify ellipses, images have to be preprocessed first by an edge detection method which produces an edge map. Then, the (x_i, y_i) pixel coordinates for each edge p_i is collected inside the edge vector $P = \{p_1, p_2, \ldots, p_{N_p}\}$, where N_p is the number of edge elements.

Each ellipse solution E (agent) employs five edge pixels. With this representation, edge pixels are selected considering a random index within P. This process will combine in a candidate solution the ellipse that hypothetically passes through five points p_1, p_2, p_3, p_4 and p_5 ($E = \{p_1, p_2, p_3, p_4, p_5\}$). Therefore, by substituting the pixel coordinates of each edge from E into Eq. 9.6, we obtain a set of five simultaneous equations that are linear in the five unknown parameters a', b', c', f' and g'.

$$a'x^2 + 2h'xy + b'y^2 + 2g'x + 2f'y + c' = 0 \qquad (9.5)$$

Then, solving the equations and dividing by the constant c', it is produced:

$$ax^2 + 2hxy + by^2 + 2gx + 2fy + 1 = 0 \qquad (9.6)$$

Considering the arrangement of the edge pixels illustrated in Fig. 9.2, the ellipse center (x_0, y_0), the maximal radius (r_{max}), the minimal radius (r_{min}) and the orientation (θ) can be computed as follows:

$$x_0 = \frac{hf - bg}{C}, \qquad (9.7)$$

$$y_0 = \frac{gh - af}{C}, \qquad (9.8)$$

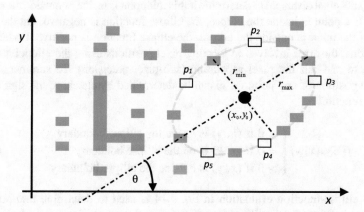

Fig. 9.2 Ellipse prototype produce from the combination of points p_1, p_2, p_3, p_4 and p_5

$$r_{\max} = \sqrt{\frac{-2\Delta}{C(a+b-R)}},$$ (9.9)

$$r_{\min} = \sqrt{\frac{-2\Delta}{C(a+b+R)}},$$ (9.10)

$$\theta = \frac{1}{2}\arctan\left(\frac{2h}{a-b}\right)$$ (9.11)

So that

$$R^2 = (a-b)^2 + 4h^2, \quad C = ab - h^2 \text{ and } \Delta = \det\left(\begin{vmatrix} a & h & g \\ h & b & f \\ g & f & 1 \end{vmatrix}\right)$$ (9.12)

Optimization corresponds to selecting the best solution from a set of available alternatives. In the simplest case, it means to obtain the optimal value of a cost function or error by systematically searching the values of variables from their valid ranges. To compute the error produced by a candidate solution E, the ellipse coordinates are calculated as a virtual shape which, in turn, must also be validated, i.e., if it really exists in the edge image. The test set is represented by $S = \{s_1, s_2, \ldots, s_{N_s}\}$, where N_s are the number of points over which the existence of an edge point, corresponding to E, should be tested.

The set S is produced considering the Midpoint Ellipse Algorithm (MEA) [24]. This scheme is a searching method that collects the required points for drawing an ellipse. Under this MEA, any pixel (x, y) on the limits of the ellipse with a, h, b, g and f solves the equation $f_{ellipse}(x, y) \cong r_x x^2 + r_y y^2 - r_x^2 r_y^2$. Nevertheless, MEA avoids calculating the square root operations by comparing the pixel distances. A method for direct distance comparison is to evaluate the halfway distance between two different pixels (sub-pixel distance) to determine if this midpoint is inside or outside the ellipse limits. If a point is inside the ellipse, the ellipse function is negative. On the other hand, if the point is outside the ellipse, the ellipse function is positive. Under such conditions, the error involved in locating pixel positions using the midpoint test is limited to one-half the pixel separation (sub-pixel precision). To summarize, the relative position of any point (x, y) can be determined by checking the sign of the ellipse function:

$$f_{Circle}(x, y)\begin{cases} < 0 \text{ if } (x, y) \text{ is inside the ellipse boundary} \\ = 0 \text{ if } (x, y) \text{ is on the ellipse boundary} \\ > 0 \text{ if } (x, y) \text{ is outside the ellipse boundary} \end{cases}$$ (9.13)

The ellipse-function evaluation in Eq. 9.14 is used to determine mid-position between pixels nearby the ellipse path at each sampling step. Figure 9.3a, b show the midpoint between two candidate pixels at a sampling position. The ellipse is used

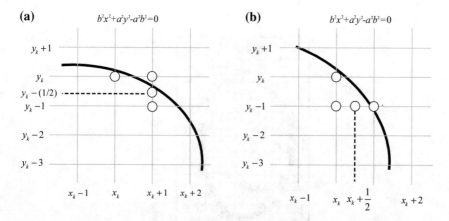

Fig. 9.3 a Symmetry of the ellipse: an estimated octant that uses the first region where the slope is greater than -1, **b** in this area, the slope presents a value less than -1

to divide the quadrants into two regions the limit of the two regions is the point at which the curve has a slope of -1 as shown in Fig. 9.3.

In MEA the time of its operation is reduced by using the symmetry of ellipses. Ellipses sections in adjacent octants within one quadrant maintain similar properties. MEA is considered the quickest method providing a sub-pixel precision [25]. Nevertheless, in order to protect the MEA operation, it is important to assure that points locating outside the image plane are not used in S.

The cost function $J(E)$ corresponds to the approximation error found among the pixels S of the ellipse candidate E and the pixels that actually exist in the edge image, yielding:

$$J(E) = 1 - \frac{\sum_{v=1}^{Ns} G(x_v, y_v)}{Ns} \qquad (9.14)$$

where $G(x_i, y_i)$ is a model that determines the pixel presence in (x_v, y_v), where $(x_v, y_v) \in S$ and N_s is the number of pixels lying on the perimeter corresponding to E currently under testing. Therefore, $G(x_v, y_v)$ is modeled as follows:

$$G(x_v, y_v) = \begin{cases} 1 & \text{if the pixel } (x_v, y_v) \text{ is an edge point} \\ 0 & \text{otherwise} \end{cases} \qquad (9.15)$$

A value of $J(E)$ close to zero represents a better resemblance. Figure 9.4 presents the procedure to determine a candidate ellipse E with its representation as a hypothetical shape S. Figure 9.4a symbolizes the original edge map, while Fig. 9.4b shows the hypothetical ellipse S corresponding to the agent $E = \{p_1, p_2, p_3, p_4, p_5\}$. In Fig. 9.4c, the hypothetical shape S is evaluated with regard to the original image, point by point, in order to obtain the similarities between virtual and edge points. The agent has been produced by using the points p_1, p_2, p_3, p_4 and p_5 which are

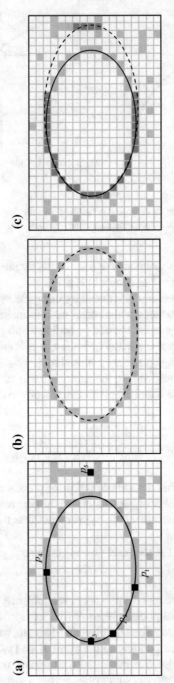

Fig. 9.4 Assessment of a candidate ellipse E: the image in **a** shows the original image while **b** presents the generated virtual shape considering the points p_1, p_2, p_3, p_4 and p_5. The image in **c** presents the resemblance between both images

illustrated in Fig. 9.4a. The hypothetical shape S, produced by MEA, collects 52 points ($N_s = 52$) with only 35 of them presented in both images (shown as darker points in Fig. 9.4c) and yielding: $\sum_{v=1}^{N_s} G(x_v, y_v) = 35$, therefore, $J(E) = 0.327$.

9.4 Ellipse Detector Process

To identify WBC, the metaheuristic detector integrates a segmentation mechanism with the ellipse metaheuristic detector presented in this chapter.

To use the detector, smear images have to be preprocessed to produce two new images: the segmented image and its corresponding edge map. The segmented image is yielded by using a segmentation method, whereas the edge map is produced by an edge detector method. This edge map is used by the cost function to evaluate the resemblance of a candidate ellipse with an actual WBC.

The objective of the segmentation scheme is to separate the WBCs from other elements such as red blood cells and background pixels. Information of color, brightness and gradients are commonly used within a thresholding value to produce the classes to discriminate each pixel. Although simple histogram thresholding can be considered to perform this operation, at this work, the Diffused Expectation-Maximization (DEM) has been used to assure better results [26].

DEM is an Expectation-Maximization (EM) based method that has been considered to segment complex medical images [27]. Different from traditional EM methods, DEM uses the spatial correlations among pixels as a part of the minimization criteria. This operation permits to segment elements in spite of noisy and complex conditions.

To segment the WBC, the DEM has been calibrated by using three classes ($K = 3$), $g(\nabla h_{ik}) = |\nabla h_{ik}|^{-9/5}$, $\lambda = 0.1$ and $m = 10$ iterations. These values correspond to the best configuration settings according to [26]. As a final result of the DEM execution, three different thresholding values are employed: the first refers to the WBC's, the second to the red blood structures, whereas the third corresponds to the background pixels. Figure 9.5b shows the segmentation results produced by the DEM scheme considering Fig. 9.5a as the original image.

Once segmented the image, then the edge map is computed. The purpose of the edge map is to reduce the image complexity but preserving the object structures. The DE-based detector works over the edge map to identify ellipsoidal objects. Several methods can be considered to extract an edge map; however, at this work, a morphological process [17] has been used to achieve this task. Morphological edge detection is a common technique to obtain borders from a binary image. Under this approach, a binary image (I_B) is eroded by a simple structure element (I_E). Then, I_B is inverted ($\overline{I}_E$) and compared with I_B ($\overline{I}_E \wedge I_B$) to obtain pixels that exist in both images. Such pixels integrate the computed edge map from I_B. Figure 9.5c presents the edge map produced by the morphological process.

The edge map is used as an input image for the ellipse detector. Table 9.1 exhibits

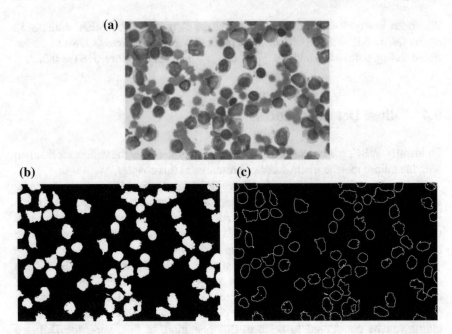

Fig. 9.5 Preprocessing process for ellipse detection. **a** original image, **b** segmented image obtained by DEM and **c** the edge map

Table 9.1 DE parameters considered in the experiments	m	F	CR	NI
	20	0.80	0.80	200

the parameter values for the DE algorithm considered in the experiments. They have obtained through exhaustive experimentation.

To present the operation of the algorithm, a numerical example has been designed to detect a single leukocyte lying inside of a simple image. Figure 9.6a presents the image considered in the example. After applying the threshold operation, the WBC is located beside a few other pixels, which are merely noise (see Fig. 9.6b). Then, the edge map is subsequently computed and stored pixel by pixel inside the vector P. Figure 9.6c shows the resulting image after such a procedure.

The DE-based ellipse detector is executed using the information of the edge map (for the sake of easiness, it only considers a population of four particles). Like all evolutionary approaches, DE is a population-based optimizer that attacks the starting point problem by sampling the search space at multiple, randomly chosen, initial particles. By taking five random pixels from vector P, four different particles are constructed. Figure 9.6) depicts the initial particle distribution $\mathbf{E}^0 = \{E_1^0, E_2^0, E_3^0, E_4^0\}$. By using the DE operators, four different trial particles $\mathbf{T} = \{T_1, T_2, T_3, T_4\}$ (ellipses) are generated, their locations are shown in Fig. 9.6e. Then, the new population $\mathbf{E}^1$ is selected considering the best elements obtained among

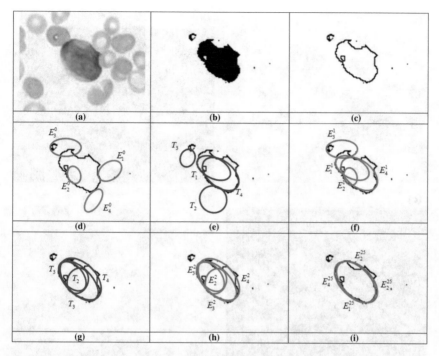

Fig. 9.6 Detection illustration: **a** Original image. **b** Segmented image. **c** Edge map. **d** Initial particles $\mathbf{E}^0$. **e** Trial elements T produced by the DE operators. **f** New population $\mathbf{E}^1$. **g** Trial elements produced considering $\mathbf{E}^1$ as input population. **h** New population $\mathbf{E}^2$. **i** Final particle configuration after 25 iterations

the trial elements **T** and the initial particles $\mathbf{E}^0$. The final distribution of the new population is depicted in Fig. 9.6). Since the particles E_2^0 and E_2^0 hold (in Fig. 9.6f) a better fitness value ($J(E_2^0)$ and $J(E_3^0)$) than the trial elements T_2 and T_3, they are considered as particles of the final population $\mathbf{E}^1$. Figures 9.6g, h present the second iteration produced by the algorithm, whereas Fig. 9.5i shows the population configuration after 25 iterations. From Fig. 9.6i, it is clear that all particles have converged to a final position which is able to accurately cover the WBC.

9.5 Experimental Results

Several experiments have been carried out to evaluate the performance of the WBC metaheuristic detector. For its evaluation, microscope images from blood-smears holding a 960×720 pixel resolution have been used. They correspond to supporting images on the leukemia diagnosis. The images present several complex contexts, such as deformed cells and overlapping with different occlusions. The robustness of the scheme has been evaluated considering complex scenarios. The robustness of

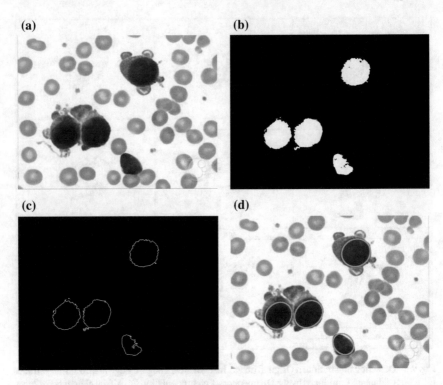

Fig. 9.7 Resulting images produced by WBC metaheuristic detector: **a** Original image, **b** image segmented, **c** edge map and **d** detected WBC

metaheuristic methods in image processing applications has been demonstrated by several interesting studies [28, 29].

Figure 9.7a presents an example considered in the study. Figure 9.7b exhibits the segmented image produced by the DEM method. Figure 9.7c, d show the edge map and the white blood cells after their identification, respectively. The results demonstrate that the metaheuristic scheme can efficiently recognize and highlight blood cells despite their complexity, deformation or overlapping. Other parameters may also be calculated through the algorithm: the total area covered by white blood cells and relationships between several cell sizes.

Another example is considered in Fig. 9.8. It shows a complex scenario with an image presenting deformed cell structures. In spite of such imperfections, the metaheuristic scheme can efficiently identify the structures, as it is shown in Fig. 9.8d.

A comprehensive set of images is considered to evaluate the performance of the metaheuristic detector. Its performance is compared with other WBC detection schemes such as the Boundary Support Vectors (BSV) approach [3], the iterative Otsu (IO) method [4], the Wang algorithm [5] and the Genetic algorithm-based (GAB) detector [16]. In all cases, the methods have been calibrated according to the values recommended by their own references.

(a)　　　　　　　　　　　　　　(b)

(c)　　　　　　　　　　　　　　(d)

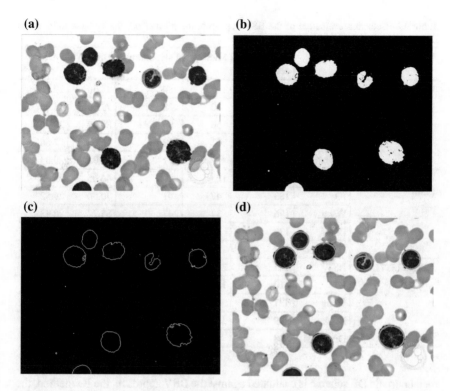

Fig. 9.8 Resulting images produced by WBC metaheuristic detector: **a** Original image, **b** segmented image, **c** edge map and **d** detected WBC

Table 9.2 reports the comparative leukocyte detection values of the BSV approach, the IO method, the Wang algorithm, the BGA detector and the metaheuristic method, in terms of false alarms and detection rates. The data set involves 50 images that have been collected from the ASH Image Bank (http://imagebank.hematology.org/). These images include 517 leukocytes (287 bright leukocytes and 230 dark leukocytes according to smear conditions), which have been identified and counted by a human observer. Such values represent the ground truth elements for all the tests. In comparison, the detection rate (DR) corresponds to the ratio between the number of leukocytes correctly detected and the number leukocytes determined by the human observer. The false alarm rate (FAR) refers to the ratio between the number of non-leukocyte objects that have been wrongly identified as leukocytes and the number leukocytes which have been actually determined by the expert. Experimental evidence demonstrates that the proposed DE method achieves 98.26% leukocyte detection accuracy with 2.71% false alarm rate.

Images of WBC are frequently contaminated by noise due to various sources of interference and other unmodeled phenomena that affect the capture process in data acquisition systems. Under such conditions, the identification results are affected by the algorithm's capacity to face several types of noise. To evaluate the robustness, the

Table 9.2 Detection evaluation of the BSV approach, the IO method, the Wang algorithm, the BGA detector and the metaheuristic DE method

Leukocyte Type	Method	Leukocytes detected	Missing	False alarms	DR (%)	FAR (%)
Bright Leukocytes (287)	BSV	130	157	84	45.30	29.27
	IO	227	60	73	79.09	25.43
	Wang	231	56	60	80.49	20.90
	BGA	220	67	22	76.65	7.66
	DE-based	281	6	11	97.91	3.83
Dark Leukocytes (230)	BSV	105	125	59	46.65	25.65
	IO	183	47	61	79.56	26.52
	Wang	196	34	47	85.22	20.43
	BGA	179	51	23	77.83	10.00
	DE-based	227	3	3	98.70	1.30
Overall (517)	BSV	235	282	143	45.45	27.66
	IO	410	107	134	79.30	25.92
	Wang	427	90	107	82.59	20.70
	BGA	399	118	45	77.18	8.70
	DE-based	508	9	14	98.26	2.71

metaheuristic DE scheme is evaluated against the BSV approach, the IO method, the Wang algorithm and the BGA detector under noisy conditions. In the experiments, two distinct tests have been analyzed. The first considers the performance of each method when the identification process is carried out over images contaminated by Salt & Pepper noise. The second test evaluates images contained Gaussian noise. Salt & Pepper and Gaussian noise are considered for the robustness analysis since they are the most popular noise commonly found in images of blood smear [30]. The test involves the whole set of 50 images presented in Sect. 6.1 containing 517 leukocytes, which have been detected and counted by a human expert. The additional noise is generated by MatLab©, assuming two noise levels of 5% and 10% for Salt & Pepper noise whereas $\sigma = 5$ and $\sigma = 10$ for the case of Gaussian noise. Figure 9.9 illustrates two different examples of the images considered in the experimental group. The results regarding the detection rate (DR) and the false alarm rate (FAR) are registered for each noise type in Tables 9.3 and 9.4. The values in the table demonstrate that the metaheuristic DE method reaches the best detection indexes, achieving in the worst case a DR of 89.55% and 91.10%, under contaminated conditions of Salt & Pepper and Gaussian noise, respectively. On the other hand, the DE detector possesses the least degradation performance presenting a FAR value of 5.99% and 6.77%.

To evaluate the stability of the metaheuristic scheme, its results are contrasted to those reported by Wang et al. in [5]. This technique is recognized as a precise method for WBC detection.

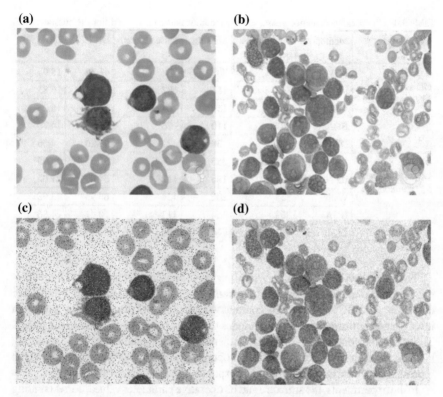

Fig. 9.9 Images considered in the comparison. **a, b** Originals images. **c** Image corrupted with 10% of Salt & Pepper noise and **d** image contaminated with $\sigma = 10$ of Gaussian noise

Table 9.3 Comparative detection results among methods under noise conditions (Salt & Pepper noise)

Noise level	Method	Leukocytes detected	Missing	False alarms	DR (%)	FAR (%)
5% Salt & Pepper noise 517 Leukocytes	BSV	185	332	133	34.74	26.76
	IO	311	206	106	63.38	24.88
	Wang	250	176	121	58.68	27.70
	BGA	298	219	135	71.83	24.18
	DE-based	482	35	32	91.55	7.04
10% Salt & Pepper noise 517 Leukocytes	BSV	105	412	157	20.31	30.37
	IO	276	241	110	53.38	21.28
	Wang	214	303	168	41.39	32.49
	BGA	337	180	98	65.18	18.95
	DE-based	463	54	31	89.55	5.99

Table 9.4 Comparative detection results among methods under noise conditions (Gaussian noise)

Noise level	Method	Leukocytes detected	Missing	False alarms	DR (%)	FAR (%)
$\sigma = 5$ Gaussian noise 517 Leukocytes	BSV	214	303	98	41.39	18.95
	IO	366	151	87	70.79	16.83
	Wang	358	159	84	69.25	16.25
	BGA	407	110	76	78.72	14.70
	DE-based	487	30	21	94.20	4.06
$\sigma = 10$ Gaussian noise 517 Leukocytes	BSV	162	355	129	31.33	24.95
	IO	331	186	112	64.02	21.66
	Wang	315	202	124	60.93	23.98
	BGA	363	154	113	70.21	21.86
	DE-based	471	46	35	91.10	6.77

The Wang method is an energy-minimizing algorithm that is operated by restrictive internal elements and guided by external gradient forces, producing the active detection of WBC's at a closed contour. As gradient values, the Wang approach involves edge information, which includes the gradient extent of pixels in the image. Therefore, the contour is attracted to elements with large gradient values, i.e., strong borders. At each step, the Wang scheme obtains a new contour configuration, which represents the minimal energy that corresponds to the gradient values.

In the experiments, the structure and its operative parameter values, corresponding to the Wang algorithm, are configurated as it is suggested in [5] while the parameters for the DE-based algorithm are taken from Table 9.1.

Figure 9.10 presents the results of both schemes assuming a test image with only two WBCs. Since the Wang method uses gradient information to correctly detect a new contour shape, it requires to be executed several iterations to identify each structure (WBC). Figure 9.10b presents the results after the Wang approach has been applied, considering only 200 iterations. Furthermore, Fig. 9.10c exhibits the identification after applying the DE-based method, which has been presented in this chapter.

The Wang algorithm employs the fuzzy cellular neural network (FCNN) as an optimization scheme. It uses gradient information and internal states to find a better contour configuration. In each step, the FCNN tests different new pixel locations which have to be located close the original border limits. This fact may cause that the contour solution remains trapped into a local minimum. To avoid such a problem, the Wang method applies a considerable number of iterations so that a near-optimal contour configuration can be found. Nevertheless, when the number of generations increases the possibility to include other elements is also high. Thus, if the image maintains a complex structure (just as smear images do) or the WBC's are overlapped, the method gets confused so that finding the correct contour configuration from the gradient magnitude is not easy. Therefore, a drawback of Wang's method is related to

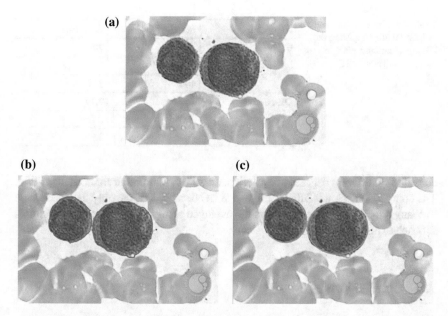

Fig. 9.10 Comparison between the DE and the Wang's method. **a** Original image. **b** Detection considering Wang's method, **c** Detection considering the DE method

its optimal iteration number (instability). Such a number must be determined experimentally as it depends on the image context and its complexity. Figure 9.11a shows the result of applying 400 cycles of Wang's algorithm while Fig. 9.11b presents the detection of the same cell shapes after 1000 iterations using the proposed algorithm. From Fig. 9.11a, it can be seen that the contour produced by Wang´s algorithm degenerates as the iteration process continues, wrongly covering other shapes lying nearby.

In order to compare the accuracy of both methods, the estimated WBC area, which has been approximated by both approaches, is compared to the actual WBC size

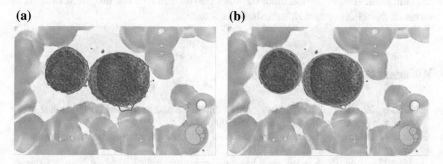

Fig. 9.11 Results for the WBC **a** Wang's algorithm after 400 iterations and **b** DE detector method considering 1000 cycles

Table 9.5 Error in the estimation after applying the DE algorithm and Wang's method to detect WBC

Algorithm	Iterations	Error (%)
Wang	60	70
	200	1
	400	121
DE-based	60	7.17
	200	2.25
	400	2.25

considering different degrees of evolution, i.e., the cycle number for each algorithm. The comparison acknowledges only one WBC because it is the only detected shape in Wang's method. Table 9.5 shows the averaged results of over twenty repetitions for each experiment.

9.6 Conclusions

In this chapter, an approach for the automatic detection of blood cell images based on the DE algorithm is presented. In the scheme, the identification process is considered as a multiple ellipse detection problem. The algorithm employs five edge points to produce candidate ellipses in the edge map of the smear. A cost function permits to accurately evaluate the similarity of a candidate ellipse with an actual WBC on the image. Guided by the values of such objective function, the set of encoded candidate ellipses are evolved using the DE algorithm so that they can fit into actual WBC on the image. The approach generates a sub-pixel detector that can effectively identify leukocytes in real images.

The performance of the DE-method has been compared to other existing WBC detectors (the Boundary Support Vectors (BSV) approach [3], the iterative Otsu (IO) method [4], the Wang algorithm [5] and the Genetic algorithm-based (GAB) detector [16] considering several images which exhibit different complexity levels. Experimental results demonstrate the high performance of the proposed method in terms of detection accuracy, robustness and stability.

References

1. Zhuang, X., Meng, Q.: Local fuzzy fractal dimension and its application in medical image processing. Artif. Intell. Med. **32**, 29–36 (2004)
2. Scholl, I., Aach, T., Deserno, T.M., Kuhlen, T.: Challenges of medical image processing. Comput. Sci Res. Dev. **26**, 5–13 (2011)
3. Wang, M., Chu, R.: A novel white blood cell detection method based on boundary support vectors. In: Proceedings of the 2009 IEEE International Conference on Systems, Man, and Cybernetics San Antonio, TX, USA - October 2009

4. Wu, J., Zeng, P., Zhou, Y., Oliver, C.: A novel color image segmentation method and its application to white blood cell image analysis. In: 8th International Conference on Signal Processing (2006)
5. Wang, S., Korris, F.L., Fu, D.: Applying the improved fuzzy cellular neural network IFCNN to white blood cell detection". Neurocomputing **70**, 1348–1359 (2007)
6. Muammar, H., Nixon, M.: Approaches to extending the Hough transform. In: Proceedings of the International Conference on Acoustics, Speech and Signal Processing ICASSP-89, vol. 3, pp. 1556–1559 (1989)
7. Atherton, T., Kerbyson, D.: Using phase to represent radius in the coherent circle Hough transform. In: IEE Colloquium on the Hough Transform, pp. 1–4. IEEE, New York (1993)
8. Díaz, P., Pérez-Cisneros, M., Cuevas, E., Hinojosa, S., Zaldivar, D.: An improved crow search algorithm applied to energy problems. Energies **11**(3), 571 (2018)
9. Shaked, D., Yaron, O., Kiryati, N.: Deriving stopping rules for the probabilistic Hough transform by sequential analysis. Comput. Vis. Image. Und. **63**, 512–526 (1996)
10. Xu, L., Oja, E., Kultanen, P.: A new curve detection method: Randomized Hough transform (RHT). Pattern Recogn. Lett. **11**(5), 331–338 (1990)
11. Han, J., Koczy, L.: Fuzzy Hough transform. In: Proceedings of the 2nd International Conference on Fuzzy Systems, vol. 2, pp. 803–808 (1993)
12. Ayala-Ramirez, V., Garcia-Capulin, C.H., Perez-Garcia, A., Sanchez-Yanez, R.E.: Circle detection on images using genetic algorithms. Pattern Recogn. Lett. **27**, 652–657 (2006)
13. Lutton, E., Martinez, P.: genetic algorithm for the detection of 2D geometric primitives in images. In: Proceedings of the 12th international conference on pattern recognition, Jerusalem, Israel, 9–13 October 1994, vol. 1, pp. 526–528 (1994)
14. Yao, J., Kharma, N., Grogono, P.: A multi-population genetic algorithm for robust and fast ellipse detection. Pattern Anal. Appl. **8**, 149–162 (2005)
15. Cheng, H.D., Guo, Y., Zhang, Y.: A novel Hough transform based on eliminating particle swarm optimization and its applications. Pattern Recogn. **42**(9), 1959–1969 (2009)
16. Karkavitsas, G., Rangoussi, M.: Object localization in medical images using genetic algorithms. World Acad. Sci. Eng. Tec. **2**, 6–9 (2005)
17. Gonzalez, R.C., Woods, R.E.: Digital Image Processing. Addison Wesley, Reading, MA (1992)
18. Babu, B., Munawar, S.: Differential evolution strategies for optimal design of shell-and-tube heat exchangers. Chem. Eng. Sci. **62**(14), 3720–3739 (2007)
19. Mayer, D., Kinghorn, B., Archer, A.: Differential evolution—An easy and efficient evolutionary algorithm for model optimization. Agr. Syst. **83**, 315–328 (2005)
20. Kannan, S., Slochanal, S.M.R., Padhy, N.: Application and comparison of metaheuristic techniques to generation expansion planning problem. IEEE Trans. Power Syst. **20**(1), 466–475 (2003)
21. Chiou, J., Chang, C., Su, C.: Variable scaling hybrid differential evolution for solving network reconfiguration of distribution systems. IEEE Trans. Power Syst. **20**(2), 668–674 (2005)
22. Cuevas, E., Zaldivar, D., Pérez-Cisneros, M.: A novel multi-threshold segmentation approach based on differential evolution optimization. Expert Syst. Appl. **37**, 5265–5271 (2010)
23. Storn, R., Price, K.: Differential evolution—A simple and efficient adaptive scheme for global optimization over continuous spaces. Technical Rep. No. TR-95–012, International Computer Science Institute, Berkley (CA) (1995)
24. Bresenham, J.E.: A linear algorithm for incremental digital display of circular arcs. Commun. ACM **20**, 100–106 (1987)
25. Van Aken, J.R.: Efficient ellipse-drawing algorithm. IEEE Comp., Graphics Appl. **4**(9), 24–35 (2005)
26. Boccignone, G., Ferraro, M., Napoletano, P.: Diffused expectation maximisation for image segmentation. Electron. Lett. **40**, 1107–1108 (2004)
27. Boccignonea, G., Napoletano, P., Caggiano, V., Ferraro, M.: A multi-resolution diffused expectation–maximization algorithm for medical image segmentation. Comput. Biol. Med. **37**, 83–96 (2007)

28. Cuevas, E.: Block-matching algorithm based on harmony search optimization for motion estimation. Appl. Intell. **39**(1), 165–183 (2013)
29. Díaz-Cortés, M.-A., Ortega-Sánchez, N., Hinojosa, S., Cuevas, E., Rojas, R., Demin, A.: A multi-level thresholding method for breast thermograms analysis using Dragonfly algorithm. Infrared Phys. Technol. **93**, 346–361 (2018)
30. Landi, G., Loli Piccolomini, E.: An efficient method for nonnegatively constrained Total Variation-based denoising of medical images corrupted by Poisson noise, Computerized Medical Imaging and Graphics, pp. 38–46 (2012)

Chapter 10
Experimental Analysis Between Exploration and Exploitation

Abstract There exist hundreds of metaheuristic methods that can be employed to obtain the optimal value in an optimization problem. To present a good performance, every metaheuristic scheme requires to achieve an adequate balance between exploration and exploitation of the search space. Even though exploration and exploitation are considered two important concepts in metaheuristics computation, the main implications with this equilibrium have not yet been completely understood. Most of the existent studies consider only the comparison of their final results, which cannot appropriately evaluate the existent balance between both concepts. This chapter conducts an experimental study where it is analyzed the balance between exploration and exploitation on several of the most popular metaheuristic schemes. In the analysis, a diversity measurement for each dimension is employed to evaluate the equilibrium of each metaheuristic approach, considering a representative set of different optimization problems.

10.1 Introduction

Recently, metaheuristics computation schemes have been considered as tools to solve a wide amount of problems in different application domains, involving digital image processing, engineering design, data analysis, computer vision, networks and communications, power, and energy management, machine learning, robotics, medical diagnosis, and others [1].

Several metaheuristic approaches emulate population-based search models, in which a group of agents (or individuals) uses particular operators to explore different possible solutions inside a solution space. This optimization methodology maintains different advantages, which include the interaction among individuals (which favors the exchange of information among different solutions) and the diversification of the agents (which is critical to enforce the efficient exploration of the search space and the capacity to avoid local optimal points) [2].

Metaheuristic computation schemes represent a great variety in terms of design and potential applications. In spite of their great number, there still exists a question

© Springer Nature Switzerland AG 2021   249
E. Cuevas et al., *Metaheuristic Computation: A Performance Perspective*,
Intelligent Systems Reference Library 195,
https://doi.org/10.1007/978-3-030-58100-8_10

which it is necessary to answer: Which element in the design of a metaheuristic scheme presents a major impact in its better performance? One commonly accepted criterion in the specialized community considers that metaheuristic schemes can produce a better performance when an adequate equilibrium between exploration and exploitation in the production of solutions [3]. While there seems to have a general agreement on this concept, in fact, there exists a vague knowledge of what the balance exploration-exploitation really refers [4, 5]. Indeed, the division of evolutionary operators and search strategies contained in a metaheuristic method is often unclear, since they can be considered of the explorative or exploitative nature [6].

Under scarce knowledge about the process that determines the balance exploration-exploitation, different trials have been carried out to fulfill these gaps. Most of these attempts have introduced interesting metrics that permit to evaluate the amount of exploration and exploitation in search strategies through the analysis of the current population diversity [4–11]. Although distinct metrics exist, there is no objective way to assess the proportion of exploration/exploitation delivered by a metaheuristic method.

One of these indexes is the evaluation of the dimension-wise diversity introduced in [10] and [12]. This metric determines the averaged distance among all the solutions to the median value of each decision variable. Then, the exploration percent at each generation is computed by considering the diversity value between the maximum diversity value encounter in the whole optimization process. Likewise, the exploitation rate is determined as the inverse of the exploration values. This measurement permits us to know how distributed or clustered. The search agents are over a particular generation during the evolution process. These numbers give knowledge of the period of time that the metaheuristic scheme behaves, exploring or exploiting solutions. Although the information provided by this index is interesting, it has not been adopted by the metaheuristic community yet to evaluate the balance between exploration and exploitation.

In this chapter, it is presented an experimental analysis to evaluates the balance between exploration and exploitation of several metaheuristic algorithms. As a result, the analysis delivers several observations that allow understanding of how this balance has an effect on the results of each type of functions, and how this balance produces better solutions. The schemes considered in this work involves schemes such as Bat Algorithm (BA), Artificial Bee Colony (ABC), Crow Search Algorithm (CSA), Covariance Matrix Adaptation Evolution Strategies (CMA-ES), Firefly Algorithm (FA), Differential Evolution (DE), Moth-Flame Optimization (MFO), Grey Wolf Optimizer (GWO), Social Spiders Optimization (SSO), Particle Swarm Optimization (PSO), Whale Optimization Algorithm (WOA) and Teaching-Learning Based Optimization (TLBO). These methods represent the most important metaheuristic search algorithms on the current literature in virtue of their performance and potential.

10.2 Exploration and Exploitation

In metaheuristic computation, exploration and exploitation correspond to the most important concept for achieving success when a particular method faces an optimization problem. Exploration represents the ability of a search strategy to produce solutions spreading in different regions of the search space. On the other hand, exploitation highlights the concept of strengthening the search process over promising regions of the solution space to locate better solutions or improve existing ones [13]. About these concepts, practical experimentation has confirmed that there is a clear relationship between the exploration-exploitation ability of a specific search technique and its convergence rate. In general, while exploitation procedures are identified to improve the convergence rate toward a global optimum, they are also recognized to raise the probability of being trapped into local optima.

Conversely, search strategies that promote exploration over exploitation conduce to rise in the probability of detecting areas of the search space where the global optimum can be located. This situation produces a deterioration of the convergence speed [14].

Recently, questions about the way in which exploration and exploitation of solutions are achieved in metaheuristic optimization have remained as an open subject. Although these questions seem trivial, their answers represent a source of disagreement among many researchers [15, 16]. Even though many answers to these questions seem opposite, there is a common agreement within the research community on the idea that a good ratio between exploration and exploitation is essential to ensure good performance in this kind of search method. Under this scenario, it is important to question: Which rate represents the optimal combination of exploration and exploitation to produce an efficient search strategy? Given that metaheuristic schemes can be very different in terms of their operators, it is complicated (if not impossible) to produce an appropriate exploration/exploitation combination that performs competitively for every optimization problem. Therefore, it is clear that the comprehension of the mechanisms implemented by these methods (and how these contribute to the search process) is necessary to devise an efficient search strategy.

Some operators, such as the selection mechanisms, allow choosing promising solutions among the available elements within the population. They are commonly employed to combine elitism and diversity [17]. In the case of greedy selection mechanisms, for example, it is promoted that the best individuals among possible solutions are unalerted for the new generation. This mechanism is used to improve convergence speed toward promising solutions. Also, in methods such as DE [18] or HS [19], where a greedy selection process is considered, new elements are selected only if they present better quality than original solutions. This selection strategy has the potential to enhance the combination of the exploration-exploitation for the production of new solutions since the best solutions, along with other elements (of low quality) can be selected to balance the rate elitism and diversity. [20].

On the other hand, there exist other metaheuristic methods that do not consider any kind of selection mechanism in their structure. They commonly accept a new solution independently of their quality. While these schemes do not consider the selection of promising solutions during their search process, the implementation of other search mechanisms is necessary to balance the exploration-exploitation rate. In the case of algorithms such as PSO [21] or GWO [22], their search mechanisms consider an attraction process in order to enhance their exploitation capabilities. These mechanisms allow producing new solutions, moving solutions of low quality toward the position of solutions with good quality. In these schemes, how the solutions are selected as attractors, and how other solutions are moved to these attractors entirely rely on the design of the search method itself. In the case of the PSO algorithm, for example, individuals are not only set to experience an attraction toward the global best solution at a given iteration (cycle) of the search process but also toward the best solution(s) recorded by each particle as the search process evolves (personal best solution). While this approach is often considered a well-balanced attraction mechanism regarding convergence and solutions diversity, implementing this kind of search strategy requires the allocation of additional memory, which could be undesired depending on the intended application(s).

Some other algorithms also use attraction processes that consider the effect of more than one attractor to produce new solutions. These group of attractors involves a subset of all currently available solutions, or even, the complete population. There are approaches that use more complex attraction mechanisms. They do not consider only very specific solutions among the available elements, but also other specific characteristics. In FA [23], for example, the attraction exerted by each individual is not only dependent on the quality of other members. Such an attraction also considers the distance that separates the solutions involved in the attraction process. Likewise, in GSA, the attraction is determined by a mechanism so-called "gravitation force" exerted among particles within the feasible search space. Under this mechanism, the magnitude of the attraction depends not only on the fitness value of each solution but also on the distance separating them [24].

Furthermore, there are metaheuristic schemes that do not consider attraction mechanisms as part of their search strategy. These methods produce new solutions by using pure random walks (as in the case of HS) or by taking other criteria into account (i.e., the distance between solutions, as in the case of DE). In schemes such as GA, different solutions are produced by combining the information of randomly chosen elements (crossover), whereas other solutions are generated by generating displacements to currently existing solutions (mutation).

Finally, it is remarkable that metaheuristic methods consider the number of iterations as part of their search strategy. This information is mostly used to adjust the exploration-exploitation rate by modifying the value of several parameters. However, as a consequence of this constant modification on the exploration-exploitation rate, the convergence of the method to a good solution may suffer a significant impact [25].

10.3   Exploration-Exploitation Evaluation

Metaheuristic methods employ a set of possible solutions to explore the search space to find satisfactory solutions for an optimization problem at hand. In general, individuals with the best qualities tend to direct the search process towards them. As a result of this mechanism, the distance among solutions decreases while the effect of exploitation increases. On the other hand, if the distance between solutions increases, the exploration effect is clearer.

To compute the variations in distance among solutions, a diversity metric known as the dimension-wise diversity index [10] is taken into consideration. Under this scheme, the population diversity is formulated as follows:

$$Div_j = \frac{1}{n} \sum_{i=1}^{n} \left| median(x^j) - x_i^j \right|$$

$$Div = \frac{1}{m} \sum_{j=1}^{m} Div_j \tag{10.1}$$

where $median(x^j)$ symbolizes the median of dimension j in the complete population. x_i^j represents the dimension j of solution i. n is the number of solutions in the population, while m corresponds to the number of decision variables of the optimization problem at hand.

The diversity of every dimension Div_j is formulated as the length between the dimension j of each solution and the median value of that dimension, averaged. The diversity of the complete population Div is then computed by using the mean value of every Div_j in each dimension. Both values are evaluated in each generation.

Once computed this information, the rate of exploration-exploitation can be evaluated as the percentage of the time that a metaheuristic scheme performs exploration or exploitation in terms of its diversity. These quantities are computed in every generation using these models:

$$XPL\% = \left(\frac{Div}{Div_{max}} \right) \times 100$$

$$XPT\% = \left(\frac{|Div - Div_{max}|}{Div_{max}} \right) \times 100 \tag{10.2}$$

where Div_{max} corresponds to the maximal diversity value obtained in the complete optimization process.

The percentage of exploration $XPL\%$ corresponds to the amount of exploration. It defines the relationship between diversity in every iteration and the maximum value of diversity. The percentage of exploitation $XPT\%$ represents the amount of exploitation. This value is computed as the complementary percentage to $XPL\%$ since the difference between the maximal diversity and the current diversity of a generation is provoked as a consequence of the concentration of candidate solutions. As can be seen, both magnitudes $XPL\%$ and $XPT\%$ are mutually conflicting and complementary.

To show the computation of these quantities, a graphical example is considered. In the example, a hypothetical metaheuristic method is employed to obtain the solution for a simple two-dimensional optimization problem defined as follows:

$$f(x_1, x_2) = 3(1 - x_1)^2 e^{-\left(x_1^2 - x_2^2\right)} - 10\left(\frac{x_1}{5} - x_1^3 - x_2^5\right) e^{\left(-x_1^2 - x_2^2\right)}$$
$$- 1/3 e^{\left(-(x_1+1)^2 - x_2^2\right)} \tag{10.3}$$

Considering the limits $-3 \leq x_1, x_2 \leq 3$, the function presents a maximal global value and two local maxima. Figure 10.1b shows a three-dimensional graphical representation of this function. A good combination of the diversification of solutions in the search space and intensification of the best-found solutions present a conflicting scenario that should be analyzed. Metaheuristic schemes present a generic design scheme, in spite of their existent differences. In the beginning, the metaheuristic

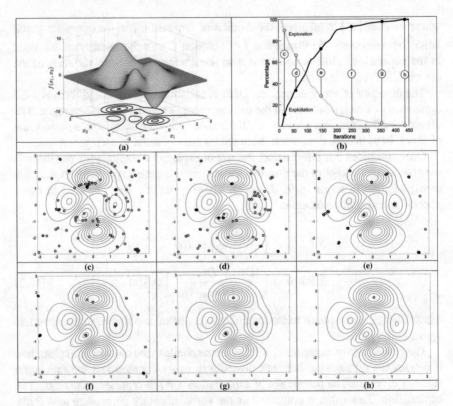

Fig. 10.1 Performance behavior during 450 iterations produced by a hypothetical method in terms of the balance evaluation defined by Eq. 10.2

method favors diversification, which means the generation of candidate solutions in different positions of the search space [26–30]. As the evolution process progress, the exploitation should be increased to refine the quality of their solutions. Figure 10.1b illustrates an example of the operation of this process during 450 generations generated by a hypothetical scheme with respect to the balance evaluation defined by Eq. 10.2. Six locations (c), (d), (e), (f), (g) and (h) have been chosen to illustrate the solution distribution and their corresponding balance evaluations. Point (c) shows an early period of the process (approximately the 20th iteration), where the balance evaluation presents the values of $XPL\% = 90$ and $XPT\% = 10$. With these percentages, the hypothetical approach operates with a tendency to the process of exploration. This situation can be shown by Fig. 10.1c in which the solutions present a wide dispersion of the search space. Point (d) represents the 60th generation. In this step, the balance value acquiree a value of $XPL\% = 70$ along with $XPT\% = 30$. Under this behavior, the metaheuristic scheme maintains mainly exploration with a small level of exploitation. The configuration of the agents under this behavior can be illustrated by Fig. 10.1d where diversity presents a high value with some groups presenting similar positions. Points (e) and (f) exemplify the 150 and 250 generations, respectively, where the balance magnitudes have as values $XPL\% = 25$, $XPT\% = 75$ and $XPL\% = 08$, $XPT\% = 92$, respectively. With such percentages, the operation of the metaheuristic scheme has been inverted, promoting more the exploitation than the exploration. Figure 10.1e, f shows the solution distribution for points (e) and (f) from Fig. 10.1b. Under such distributions, the solutions are arranged in different clusters decreasing the total diversity. Finally, points (g) y (h) correspond to the final stages of the operation of the hypothetical metaheuristic method. In such locations, the method presents an evident trend to the exploitation of the best-found solutions without taken into account any exploration process. Figure 10.1g, h represent the solution configuration for points (g) and (h).

To test that this balance evaluation does not present a bias towards exploration or exploitation, an experiment is carried out. For the evaluation, it has been was used the Random Search algorithm, which produces a random population in each iteration with a fixed standard deviation. This allowed us to see how the algorithm behaves when a fixed diversity is considered. Table 10.1 presents the maximum diversity, mean diversity and exploration and exploitation rate from the Random Search algorithm considering a standard deviation of 2.88 for each random movement and 1000 iterations. Figure 10.2 shows the exploration-exploitation rate.

A close inspection of Fig. 10.2 demonstrates how a fixed standard deviation of the algorithm affected the exploration-exploitation rate through the operation of the method. The distance between search agents is always within the limits of the standard deviation. As the diversity in each iteration is always near to the maximal diversity,

Table 10.1 Results of the experiment that considers the Random Search algorithm in a control test

Div_{max}	Div_{mean}	%XPL	%XPT
2.57E+00	2.45E+00	95.52	4.48

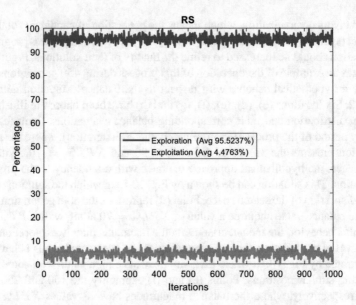

Fig. 10.2   Exploration-exploitation rate of the random search algorithm

an average exploration rate of 95.52 is obtained. From this experiment, it is clear
that the algorithm presents a tendency to explore all the time while the exploitation
is very small. This behavior is congruent with the random search algorithm since it
does not contain any exploitative operator.

10.4   Experimental Results

To evaluate the rate of exploration-exploitation in several of the most popular metaheuristic algorithms, it is studied the performance of 8 state-of-the-art optimization
techniques: Covariance Matrix Adaptation Evolution Strategy (CMA-ES) [25], Artificial Bee Colony (ABC) [22], [23], Differential Evolution (DE) [16], Crow Search
Algorithm (CSA) [26], Moth Flame Optimization (MFO) [13], Grey Wolf Optimization (GWO) [31] and Whale Optimization Algorithm (WOA) [32]. The algorithms
used in our analysis have been selected to consider a wide range of search mechanisms and design methodologies. Another reason has been to involve recognized
schemes and newly proposed methods. The availability of source code has also been
an important factor, as well as the popularity and performance reported in the literature. The analysis has been conducted considering 10 well-known benchmark test
functions. Table 10.2 shows the set of used functions.

The algorithms used in this study have been configured following as a guideline
their own references. Therefore, each metaheuristic methods present the calibration

Table 10.2 Functions employed to compute the exploration-exploitation rates

f_i	Name	Function	S	Dim	Minimum
f_1	Ackley	$f(x) = -20e^{-0.2\sqrt{\frac{1}{n}\sum_{i=1}^{n}x_i^2}} - e^{\frac{1}{n}\sum_{i=1}^{n}\cos(2\pi x_i)} + 20 + e$	$[-30, 30]^n$	$n = 30$	$f(\mathbf{x}^*) = 0;$ $\mathbf{x}^* = (0, \ldots, 0)$
f_2	Dixon Price	$f(x) = (x_i - 1)^2 + \sum_{i=2}^{n} i\left(2x_i^2 - x_{i-1}\right)^2$	$[-10, 10]^n$	$n = 30$	$f(\mathbf{x}^*) = 0;$ $\mathbf{x}^* = 2^{-\frac{2^i-2}{2^i}}$ for $i = 1, \ldots, n$
f_3	Griewank	$f(x) = \frac{1}{4000}\sum_{i=1}^{n}x_i^2 - \prod_{i=1}^{n}\cos\left(\frac{x_i}{\sqrt{i}}\right) + 1$	$[-600, 600]^n$	$n = 30$	$f(\mathbf{x}^*) = 0;$ $\mathbf{x}^* = (0, \ldots, 0)$
f_4	Infinity	$f(x) = \sum_{i=1}^{n}x_i^6\left[\sin\left(\frac{1}{x_i}\right) + 2\right]$	$[-1, 1]^n$	$n = 30$	$f(\mathbf{x}^*) = 0;$ $\mathbf{x}^* = (0, \ldots, 0)$
f_5	Levy	$f(x) = \cos 2(\pi w_1) + \sum_{i=1}^{n-1}(w_i - 1)^2(1 + 10\sin\pi w_i + 1)$ $+ (w_n - 1)^2(1 + \sin^2 2\pi w_n)$ $w_i = 1 + \left(\frac{x_i + 1}{4}\right)$	$[-10, 10]^n$	$n = 30$	$f(\mathbf{x}^*) = 0;$ $\mathbf{x}^* = (1, \ldots, 1)$
f_6	Mishra11	$f(\mathbf{x}) = \left[\frac{1}{n}\sum_{i=1}^{n}|x_i| - \left(\prod_{i=1}^{n}|x_i|\right)^{\frac{1}{n}}\right]^2$	$[-10, 10]^n$	$n = 30$	$f(\mathbf{x}^*) = 0;$ $\mathbf{x}^* = (0, \ldots, 0)$

(continued)

Table 10.2 (continued)

f_i	Name	Function	S	Dim	Minimum
f_7	Multimodal	$f(x) = \sum_i^n |x_i| \times \prod_i^n |x_i|$	$[-10, 10]^n$	$n = 30$	$f(\mathbf{x}^*) = 0;$ $\mathbf{x}^* = (0, \dots, 0)$
f_8	Penalty1	$f(x) = \dfrac{\pi}{30} \left\{ \begin{array}{l} 10\sin^2(\pi y_1) \\ + \sum_{i=1}^{n-1} (y_i - 1)^2 \big[1 + 10\sin^2(\pi y_i + 1)\big] \\ + (y_n - 1)^2 \end{array} \right\}$ $+ \sum_{i=1}^{n} u(x_i, 10, 100, 4);$ $y_i = 1 + \dfrac{x_i + 1}{4};$ $u(x_i, a, k, m) = \begin{cases} k(x_i - a)^m, & x_i > a \\ 0, & -a \leq x_i \leq a \\ k(-x_i - a)^m, & x_i < -a \end{cases}$	$[-50, 50]^n$	$n = 30$	$f(\mathbf{x}^*) = 0;$ $\mathbf{x}^* = (-1, \dots, -1)$

(continued)

Table 10.2 (continued)

f_i	Name	Function	S	Dim	Minimum
f_9	Penalty2	$f(x) = 0.1\left\{\sin^2(3\pi x_1) + \sum_{i=1}^{n-1}(x_i-1)^2[1+\sin^2(3\pi x_{i+1})] + (x_n-1)^2[1+\sin^2(2\pi x_n)]\right\}$ $+ \sum_{i=1}^n u(x_i, 5, 100, 4);$ $u(x_i, a, k, m) = \begin{cases} k(x_i-a)^m, & x_i > a \\ 0, & -a \le x_i \le a \\ k(-x_i-a)^m, & x_i < -a \end{cases}$	$[-50, 50]^n$	$n = 30$	$f(\mathbf{x}^*) = 0$; $\mathbf{x}^* = (1,\ldots,1)$
f_{10}	Perm1	$f(x) = \sum_{k=1}^n \left\{\sum_j^n (j^k + \beta)\left[\left(\frac{x_j}{\beta}\right)^2 - 1\right]\right\}^2$	$[-n, n]^n$	$n = 30$	$f(\mathbf{x}^*) = 0$; $\mathbf{x}^* = (1, 2, \ldots, n)$

Table 10.3 Parameter configuration for every metaheuristic scheme

Algorithm	Parameters
ABC	Population size $= 124$, Onlooker bees $= 62$, Employed bees $= 62$, Scout bee $= 1$, limit $= 100$ [23]
CMA-ES	The method has been set considering the guidelines provided by its reference [37]
CSA	Population size $= 50$, Flight length $= 2$, awareness probability $= 0.1$ [30]
DE	The weight factor is configurated to $F = 0.75$, while the crossover probability is configurated to $CR = 0.2$ [REF].
GWO	Population size $= 30$ [32]
HS	The Harmony Memory Rate is set to HCMR $= 0.7$, while the pitch adjustment rate is configurated to $PA = 0.3$ [39]
MFO	The number of flames is set as $N_{flames} = round\left(\left(N_{pop} - k\right) * \frac{N_{pop}-1}{k_{max}}\right)$ where N_{pop} denotes the population size, k the current iteration and k_{max} the maximum number of iterations [34]
PSO	The learning factors are set to $c_1 = 2$ and $c_2 = 2$. On the other hand, the inertia weight factor is configurated to decrease linearly from 0.9 to 0.2 as the process evolves [40]
WOA	Whale population $= 30$ [36]

shown in Table 10.3. The values of these parameters represent the best possible performance obtained by each metaheuristic scheme.

In all algorithms, the maximum number of function accesses has been set to 50,000 and unless otherwise stated, the population is set to 50. This values have been assumed in order to maintain compatibility with other works reported in the literature [30, 32–40]. In the following results, five different metrics have been considered: The average, median and standard deviation of the best-found solutions from 30 independent executions (**AB, MD, SD**), the average percentage of time spent exploring (**%XPL**) and the average percentage of time spent exploiting (**%XPT**).

Table 10.4 exhibits a comparison of the considered algorithms for multimodal functions. The best-obtained results are highlighted in boldface. A close inspection of Table 10.4 demonstrates that the best performing methods are the CMAES and WOA.

Figure 10.3 presents the progress of the balance of exploration-exploitation obtained by the best three algorithms for the ten multimodal functions considering all iterations. CMA-ES presents the best performance in comparison with the other schemes. It exploited the search space about 96.0574% of the time and explored 3.9426% of the time. TLBO won in 2 functions and tied in 1 with average exploitation of 91.7911% and 8.2089% exploration. Finally, WOA found the best solution on 2 functions and tied in 3; this algorithm employed a balance of 93.4241% exploitation and 6.5759% exploration. These results suggest that the best balance for multimodal functions is closer to 90% exploitation and 10% exploration. These balances of exploration-exploitation have been produced as a consequence of the search mechanisms employed by each metaheuristic scheme. CMA-ES and WOA consider a

Table 10.4 Best solutions found and rated produced by each multimodal function

		ABC	CMAES	CSA	DE	GWO	MFO	PSO	WOA
f_1	AB	1.72E+01	2.06E+00	1.71E+01	5.12E−03	1.01E−14	9.22E+00	5.44E+00	**3.08E−15**
	MD	1.73E+01	7.11E−15	1.72E+01	4.84E−03	8.88E−15	1.39E+01	2.38E−02	**3.55E−15**
	SD	4.69E−01	6.30E+00	4.15E−01	9.34E−04	3.24E−15	8.37E+00	7.98E+00	2.59E−15
	%XPL	87.4970	2.0427	98.0255	13.4553	0.4716	6.7257	44.4648	1.8393
	%XPT	12.5030	97.9573	1.9745	86.5447	99.5284	93.2743	55.5352	98.1607
f_2	AB	2.86E+05	6.67E−01	1.69E+05	1.46E+00	**6.67E−01**	4.28E+04	1.80E+04	**6.67E−01**
	MD	2.74E+05	6.67E−01	1.70E+05	1.38E+00	**6.67E−01**	8.24E+01	3.21E+02	**6.67E−01**
	SD	7.76E+04	6.03E−15	4.70E+04	3.65E−01	4.94E−06	1.16E−05	4.27E+04	8.19E−05
	%XPL	93.9632	2.4524	98.5658	15.3386	0.5150	5.7613	45.8991	2.5674
	%XPT	6.0368	97.5476	1.4342	84.6614	99.4850	94.2387	54.1009	97.4326
f_3	AB	1.37E+02	1.31E−03	1.70E+02	6.01E−03	1.05E−03	1.19E−02	9.03E+00	1.34E−03
	MD	1.39E+02	0.00E+00	1.73E+02	3.98E−03	0.00E+00	9.87E−03	8.86E−03	0.00E+00
	SD	1.61E+01	3.47E−03	2.03E+01	5.66E−03	4.40E−03	1.32E−02	2.75E+01	7.35E−03
	%XPL	78.6281	2.0703	98.3754	9.9041	0.4640	4.9714	38.7490	1.7932
	%XPT	21.3719	97.9297	1.6246	90.0959	99.5360	95.0286	61.2510	98.2068
f_4	AB	4.98E−01	5.29E−111	1.18E−01	2.42E−16	4.92E−217	4.53E−08	2.17E−10	**0.00E+00**
	MD	4.96E−01	1.04E−112	1.07E−01	2.12E−16	1.26E−221	1.19E−09	7.22E−12	**0.00E+00**
	SD	1.56E−01	1.96E−110	4.14E−02	1.68E−16	0.00E+00	1.23E−07	9.22E−10	**0.00E+00**
	%XPL	96.9063	2.4101	98.1739	15.5385	0.5353	7.6471	49.5202	2.3214
	%XPT	3.0937	97.5899	1.8261	84.4615	99.4647	92.3529	50.4798	97.6786
	AB	8.36E+01	2.12E−01	5.83E+01	**2.52E−05**	1.11E+00	2.48E−01	4.94E+00	9.66E−02

(continued)

Table 10.4 (continued)

		ABC	CMAES	CSA	DE	GWO	MFO	PSO	WOA
f_5	MD	8.47E+01	1.50E−32	5.84E+01	**2.43E−05**	1.09E+00	2.33E−01	3.50E+00	9.51E−02
	SD	1.21E+01	5.06E−01	8.60E+00	**8.90E−06**	2.14E−01	1.01E+01	4.94E+00	1.02E−01
	%XPL	95.0028	2.6241	98.1753	17.3320	2.6767	5.0162	47.1323	6.3444
	%XPT	4.9972	97.3759	1.8247	82.6680	97.3233	94.9838	52.8677	93.6556
	AB	1.91E−01	5.57E−27	9.36E−02	**0.00E+00**	7.04E−12	7.12E−07	0.00E+00	**0.00E+00**
f_6	MD	1.88E−01	0.00E+00	9.09E−02	**0.00E+00**	3.69E−12	0.00E+00	0.00E+00	**0.00E+00**
	SD	4.67E−02	3.05E−26	2.18E−02	**0.00E+00**	7.35E−12	3.90E−06	0.00E+00	**0.00E+00**
	%XPL	95.9426	2.9087	98.4887	70.1519	25.8766	4.0018	0.5549	24.2250
	%XPT	4.0574	97.0913	1.5113	29.8481	74.1234	95.9982	99.4451	75.7750
	AB	3.79E+04	1.63E−87	5.81E+01	1.69E−111	**0.00E+00**	4.66E−206	7.25E−33	**0.00E+00**
f_7	MD	5.61E+03	1.80E−121	1.14E+01	1.17E−116	**0.00E+00**	8.15E−225	1.36E−90	**0.00E+00**
	SD	1.10E+05	6.12E−87	1.24E+02	8.88E−111	**0.00E+00**	0.00E+00	3.97E−32	**0.00E+00**
	%XPL	86.4106	3.5939	98.0479	12.4436	0.6234	6.7044	37.5761	1.6233
	%XPT	13.5894	96.4061	1.9521	87.5564	99.3766	93.2956	62.4239	98.3767
	AB	1.15E+08	**7.69E+01**	4.59E+07	2.00E+05	2.27E+06	8.66E+01	9.20E+02	4.52E+02
f_8	MD	1.16E+08	**7.65E+01**	4.51E+07	1.97E+05	2.00E+06	8.73E+01	3.23E+02	1.45E+02
	SD	3.13E+07	**3.07E+00**	9.43E+06	6.77E+04	1.44E+06	3.74E+00	1.60E+03	1.19E+03
	%XPL	95.5115	2.4201	98.5632	39.5521	7.3906	6.4148	42.9836	10.5891
	%XPT	4.4885	97.5799	1.4368	60.4479	92.6094	93.5852	57.0164	89.4109
	AB	2.09E+08	1.04E+02	7.46E+07	1.92E+04	1.40E+05	**9.14E+01**	7.75E+02	4.60E+03
f_9	MD	2.08E+08	1.04E+02	7.82E+07	1.75E+04	1.25E+05	**9.18E+01**	2.38E+02	1.64E+02

(continued)

Table 10.4 (continued)

		ABC	CMAES	CSA	DE	GWO	MFO	PSO	WOA
	SD	6.05E+07	4.61E+00	1.87E+07	7.84E+03	7.09E+04	**3.72E+00**	1.54E+03	1.95E+04
	%XPL	95.3091	2.3544	99.0277	24.8489	3.7775	6.7386	48.9883	6.8941
	%XPT	4.6909	97.6456	0.9723	75.1511	96.2225	93.2614	51.0117	93.1059
	AB	6.18E+84	1.89E+83	5.10E+83	3.06E+81	3.60E+81	**1.66E+81**	4.40E+81	1.29E+82
f_{10}	MD	3.86E+84	1.02E+83	4.32E+83	1.25E+81	7.87E+80	**2.36E+80**	2.47E+81	3.02E+81
	SD	6.29E+84	2.81E+83	3.87E+83	6.49E+81	6.65E+81	**3.25E+81**	5.29E+81	2.56E+82
	%XPL	97.9017	29.1786	99.1435	89.9032	14.7970	8.4823	60.2020	12.5809

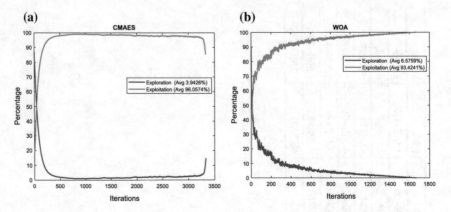

Fig. 10.3 Average balance employed by the best two algorithms in multimodal functions

non-greedy selection mechanism that provides a low exploration level. Conversely, TLBO used a greedy selection mechanism, which slightly decreases the exploration rate. On the other hand, most of them use attraction operators toward the best global solution as a search strategy. Under such conditions, they promote exploitation at a high level.

There were four outliers where CMA-ES, TLBO and WOA did not find the best solutions, functions f_5, f_9 and f_{10}. In these functions, the algorithms that obtained the best results were DE (f_5, f_7) and MFO (f_9, f_{10}). Figure 10.4 presents the evolution of the balance in these functions by the algorithms. Beginning with f_5 we can already see that DE method uses less on exploitation compared to the best algorithms, and unlike in all the other multimodal functions, the best solution in f_9 was obtained by DE with a focus on exploration, with a balance of 83.7028% of the time exploring and 16.2972% exploiting. To obtain these balance values, DE uses a search strategy that combines an increase of the exploration rate with the independence of a specific point attractor. Through these search mechanisms, DE can scape of local minima and find better solutions compared to the other metaheuristic algorithms. Another remarkable aspect of DE is the high differences produced in its balances when it faces optimization problems. On the other hand, MFO obtains the same balance level as the best three algorithms. This fact is an effect of the multiple attractors considered in its search strategy. Since it uses several attractor points, its level of exploitation decreases, incrementing the exploration of the search space slightly. This behavior is shown in Fig. 10.4 where multiple exploration peaks appear along the optimization process. The multiple attraction points allow us to jump in different zones even though it is continuously focusing on exploitation. This operation eventually permits us to find better solutions than other methods.

Figure 10.5 shows the balance levels of the worst-performing algorithms ABC and CSA. According to Fig. 10.5, it is clear that both schemes use excessive exploration in their search processes. This lack of balance results in a worse performance. It is not evident a direct relationship of this bad performance with their search mechanisms.

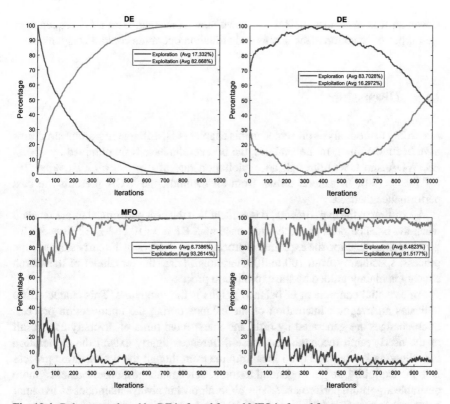

Fig. 10.4 Balance employed by DE in f_5 and f_6, and MFO in f_9 and f_{10}

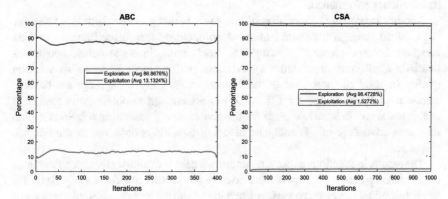

Fig. 10.5 Average balance employed by the bottom two algorithms in multimodal functions

However, it is clear that their well-known slow convergence as a consequence of their types of selection mechanisms and attraction operators could be responsible.

10.5 Discussion

According to the analysis presented in this chapter [41], different important structures have been identified. In the analysis, the best results have been obtained by CMA-ES, WOA and GWO through an exploitation rate of over 90% ($XPL\% = 10$, $XPT\% = 90$). In general terms, from all methods, the WOA presents the best performance metrics.

According to the experiments, it is evident that the best performances are reached when the balance rates produce a response of $XPL\% = 10$ and $XPT\% = 90$. To generate this behavior, the exploration process should last only the early stages of the process, specifically from 100 to 300 iterations. Then, the remainder of the search strategy is mainly guided by the exploitation process.

An essential component of balance graphs is the roughness. This characteristic indicates an irregular fluctuation of the balance during the optimization process. Such changes are generated by little and unexpected turns of diversity as a result of the used search mechanisms. These differences lightly extend the exploration rate allowing it to escape from local minima even though the optimization process is in the exploiting stage. In general terms, rough balance responses present a more desirable algorithm performance. The WOA algorithm always maintains an irregular balance response (Figs. 10.3, 10.4 and 10.5), which looks to be one of the causes of its excellent performance.

Another interesting observation of this study is that good performance behavior is produced through the combination of competitive search mechanisms and an adequate balance response. During the experiments, in some cases, two meta-heuristic algorithms presented very different performances in terms of solution quality in spite of their similar balance response. Only an appropriate balance response ($XPL\% = 10$, $XPT\% = 90$) is not enough to obtain good results. It is also necessary to have operators for the generation of promising solutions so that they take advantage of the adequate diversity conditions observed in the balance response.

The results in functions express an important case to examine since many methods could not keep the excellent performance obtained in other functions. Only CMA-ES has achieved its competitive performance in the shifted and non-shifted versions of the best performing algorithms. GWO and WOA presented a bad performance in terms of the quality solution and an inappropriate balance response. Due to its search strategy, it is well-known that GWO and WOA maintain a good performance in problems where the optimal point is in origin. However, they present severe defects when they face test functions with a shifted optimal location. Under such conditions, when an algorithm cannot determine a promising area to exploit, it requires more time to explore the search space producing an unbalance in the rate exploration-exploitation [42]. This fact affects the quality or precision of the final solutions.

10.6   Conclusion

In this paper, an empirical evaluation of the balance between exploration and exploitation on metaheuristic algorithms has been conducted. In the study, a dimension-wise diversity measurement is used to assess the balance of each scheme considering a representative set of 10 benchmark problems that involve multimodal, unimodal, composite and shifted functions. In the majority of the 10 functions (multimodal, unimodal, hybrid and shifted) the balance that produced the best results was above 90% exploitation and less than 10% exploration.

In shifted functions specifically, many algorithms maintain difficulties in maintaining the same performance that they had on to the non-shifted versions. Some of them had also problems to obtain the same balance between exploration and exploitation.

It has been observed in this study that good performance behavior is produced through the combination of competitive search mechanisms and an adequate balance response. Only an appropriate balance response is not enough to obtain good results. It is also necessary to have operators for the generation of promising solutions so that they take advantage of the adequate diversity conditions observed in the balance response.

According to this analysis, it can be formulated as a future work that the metaheuristic schemes could improve their results by gradually reducing the number of their search agents.

In general, a metaheuristic algorithm with N search agents invests N function evaluations in each iteration. In a competitive metaheuristic method, its exploitation phase lasts 90% of its execution. In this phase, the diversity of the population is small since most of the solutions are grouped or clustered in the best positions of the search space. Under such conditions, it is better to consider in the search strategy only the best B solutions from the existent N ($B < <N$). Therefore, in some cases, the computational complexity can be reduced.

References

1. Hussain, K., Mohd Salleh, M.N., Cheng, S., Shi, Y.: Metaheuristic research: a comprehensive survey. Artif. Intell. Rev. pp. 1–43 (2018)
2. Sörensen, K.: Metaheuristics-the metaphor exposed. Int. Trans. Oper. Res. **22**(1), 3–18 (2015)
3. Xu, J., Zhang, J.: Exploration-exploitation tradeoffs in metaheuristics: Survey and analysis. In: Proceedings of the 33rd Chinese Control Conference, CCC 2014, pp. 8633–8638 (2014)
4. Črepinšek, M., Liu, S.-H., Mernik, M.: Exploration and exploitation in evolutionary algorithms. ACM Comput. Surv. **45**(3), 1–33 (2013)

5. Chen, S., Boluf, A., Montgomery, J., Hendtlass, T.: An analysis on the effect of selection on exploration in particle swarm optimization and differential evolution, pp. 3037–3044 (2019)
6. Fausto, F., Reyna-Orta, A., Cuevas, E., Andrade, Á.G., Perez-Cisneros, M.: From Ants to Whales: Metaheuristics for all Tastes. Springer, The Netherlands (2019)
7. Yang, X.-S.: Swarm-based metaheuristic algorithms and no-free-lunch theorems. Theory New Appl. Swarm Intell. (2012)
8. Yang, X.-S.: Nature-inspired mateheuristic algorithms: success and new challenges. J. Comput. Eng. Inf. Technol. **01**(01) (2012)
9. Kriegel, H.P., Schubert, E., Zimek, A.: The (black) art of runtime evaluation: Are we comparing algorithms or implementations? Knowl. Inf. Syst. **52**(2), 341–378 (2017)
10. Cheng, S., Shi, Y., Qin, Q., Zhang, Q., Bai, R.: Population diversity maintenance in brain **4**(2), 83–97 (2015)
11. Al-Quraishi, T., Abawajy, J.H., Chowdhury, M.U., Rajasegarar, S., Abdalrada, A.S.: Recent advances on soft computing and data mining **700**, SCDM. Cham: Springer International Publishing (2018)
12. Salleh, M.N.M. et al.: Exploration and exploitation measurement in swarm-based metaheuristic algorithms: an empirical analysis. Adv. Intell. Syst. Comput. **700**, 24–32 (2018)
13. Yang, X.S., Deb, S., Fong, S.: Metaheuristic algorithms: Optimal balance of intensification and diversification. Appl. Math. Inf. Sci. **8**(3), 977–983 (2014)
14. Yang, X.S., Deb, S., Hanne, T., He, X.: Attraction and diffusion in nature-inspired optimization algorithms. Neural Comput. Appl. **19** (2015)
15. Díaz, P., Pérez-Cisneros, M., Cuevas, E., Hinojosa, S., Zaldívar, D.: An improved crow search algorithm applied to energy problems. Energies **11**(3), 571 (2018)
16. Storn, R., Price, K.: Differential evolution—A simple and efficient heuristic for global optimization over continuous spaces. J. Glob. Optim. **11**(4), 341–359 (1997)
17. Huang, T., Zhan, Z.-H., Jia, X., Yuan, H., Jiang, J., Zhang, J.: Niching community based differential evolution for multimodal optimization problems. In: 2017 IEEE Symposium Series on Computational Intelligence (SSCI), pp. 1–8 (2017)
18. Poli, R., Kennedy, J., Blackwell, T.: Particle swarm optimization. Swarm Intell. **1**(1), 33–57 (2007)
19. Cuevas, E.: Block-matching algorithm based on harmony search optimization for motion estimation. Appl. Intell. **39**(1), 165–183 (2013)
20. Yang, X.S., He, X.: Firefly algorithm: recent advances and applications. Int. J. Swarm Intell. **1**(1), 1–14 (2013)
21. Rashedi, E., Nezamabadi-pour, H., Saryazdi, S.: GSA: a gravitational search algorithm. Inf. Sci. (Ny) **179**(13), 2232–2248 (2009)
22. Karaboga, D.: An idea based on Honey Bee Swarm for Numerical Optimization. Tech. Rep. TR06, Erciyes Univ., no. TR06, p. 10 (2005)
23. Karaboga, D., Basturk, B.: A powerful and efficient algorithm for numerical function optimization: Artificial bee colony (ABC) algorithm. J. Glob. Optim. **39**(3), 459–471 (2007)
24. Yang, X.-S.: A new metaheuristic bat-inspired algorithm, pp. 65–74 (2010)
25. Hansen, N., Ostermeier, A.: Completely derandomized self-adaptation in evolution strategies. Evol. Comput. **9**(2), 159–195 (2001)
26. Askarzadeh, A.: A novel metaheuristic method for solving constrained engineering optimization problems: Crow search algorithm. Comput. Struct. **169**, 1–12 (2016)
27. Yang, X.-S.: Firefly algorithms for multimodal optimization. In: Lecture Notes in Computer Science (including subseries Lecture Notes in Artificial Intelligence and Lecture Notes in Bioinformatics), vol. 5792 LNCS, pp. 169–178 (2009)
28. Cuevas, E., Cienfuegos, M., Zaldívar, D., Pérez-cisneros, M.: A swarm optimization algorithm inspired in the behavior of the social-spider. Expert Syst. Appl. **40**(16), 6374–6384 (2013)
29. Rao, R.V., Savsani, V.J., Vakharia, D.P.: Teaching-Learning-Based Optimization: An optimization method for continuous non-linear large scale problems. Inf. Sci. (Ny) **183**(1), 1–15 (2012)

30. Mirjalili, S.: Moth-flame optimization algorithm: A novel nature-inspired heuristic paradigm. Knowl.-Based Syst. **89**, 228–249 (2015)
31. Mirjalili, S., Mirjalili, S.M., Lewis, A.: Grey Wolf optimizer. Adv. Eng. Softw. **69**, 46–61 (2014)
32. Mirjalili, S., Lewis, A.: The whale optimization algorithm. Adv. Eng. Softw. **95**, 51–67 (2016)
33. Chopard, B., Tomassini, M.: Particle swarm optimization. Nat. Comput. Ser. pp. 97–102 (2018)
34. Derrac, J., García, S., Molina, D., Herrera, F.: A practical tutorial on the use of nonparametric statistical tests as a methodology for comparing evolutionary and swarm intelligence algorithms. Swarm Evol. Comput. **1**(1), 3–18 (2011)
35. Hansen, N.: The CMA evolution strategy: a tutorial **102**(2006), pp. 75–102 (2016)
36. Das, S., Suganthan, P.N.: Differential evolution: A survey of the state-of-the-art. IEEE Trans. Evol. Comput. **15**(1), 4–31 (2011)
37. Fister, I., Yang, X.S., Brest, J.: A comprehensive review of firefly algorithms. Swarm Evol. Comput. **13**, 34–46 (2013)
38. Marini, F., Walczak, B.: Particle swarm optimization (PSO). A tutorial. Chemom. Intell. Lab. Syst. **149**, 153–165 (2015)
39. Awad, N.H., Ali, M.Z., Liang, J.J., Qu, B.Y., Suganthan, P.N.: Problem definitions and evaluation criteria for the CEC 2017 special session and competition on single objective real-parameter numerical optimization (2016)
40. Mania, H., Guy, A., Recht, B.: Simple random search provides a competitive approach to reinforcement learning, pp. 1–22 (2018)
41. Morales-Castañeda, B., Zaldívar, D., Cuevas, E., Fausto, F., Rodríguez, A.: A better balance in metaheuristic algorithms: Does it exist? Swarm Evol. Comput. **54**, 100671 (2020)
42. Díaz-Cortés, M.-A., Ortega-Sánchez, N., Hinojosa, S., Cuevas, E., Rojas, R., Demin, A.: A multi-level thresholding method for breast thermograms analysis using Dragonfly algorithm. Infrared Phys. Technol. **93**, 346–361 (2018)

Printed in the United States
by Baker & Taylor Publisher Services